Geometry Textbook for Middle and High School Students

by 70 Times 7 Math

Published by 70 Times 7 Math (A division of Habakkuk Educational Materials)

Copyright © 2022-2023 by 70 Times 7 Math. All rights reserved.

70 TIMES 7 MATH: GEOMETRY TEXTBOOK FOR MIDDLE AND HIGH SCHOOL STUDENTS

Copyright © 2022 by 70 Times 7 Math

All rights reserved. No part of this book may be reproduced in any form or by any electronic or mechanical means, including information storage and retrieval systems, without the written consent of the publisher. Please address your inquiries to Habakkuk@cox.net.

ISBN (Hardback Edition): 978-1-954796-60-7
ISBN (Paperback Edition): 978-1-954796-34-8

Image on the front cover and title page: © [Ihor] / Adobe Stock and 70 Times 7 Math

Certain copyrighted images in the interior of this book are used under license from stock.adobe.com. These include the classroom illustrations from the book cover images on pages 7 and 13; the cone by anatolir and the sphere by Alla on page 37; the girl illustration by yusufdemirci on page 120; the measuring tape by Marina on page 234; and the toad picture by Aleksei on page 238.

Printed and bound in the United States of America

Published by 70 Times 7 Math
(A division of Habakkuk Educational Materials)

Visit www.habakkuk.net

Table of Contents

CLASS SUPPLIES LIST FOR STUDENTS 7

OVERVIEW OF THE 70 TIMES 7 MATH CURRICULUM 8
- Chart showing supplementary student/teacher materials available for this textbook 9
- Description of the 70 Times 7 Math materials 9
- Schedule for proceeding through the material 13
- Computer-based tests for the "70 Times 7 Math Curriculum" 14

TEACHER DIRECTIONS 16

1. **CHAPTER 1 (PLANE AND 3D GEOMETRY)** 17
   - Point, line, plane, collinear, coplanar, space 18
   - Lines (Intersecting, perpendicular, concurrent, parallel lines and planes, skew lines) 19
   - Line segments, lines, rays 21
   - Curves (Curve, simple curve, closed curve, simple closed curve, polygonal curve, complex polygon, simple polygon) 22

   Polygons
   - Polygons, regular polygons, irregular polygons, equilateral, equiangular 23
   - "The Polygon Song" 24
   - Polygons (n-gon, triangle, quadrilateral, pentagon, hexagon, heptagon, octagon, nonagon, decagon, hendecagon, dodecagon) (Vocabulary: sides, angles, vertices) 25
   - Quadrilaterals (Trapezoid, rectangle, square, parallelogram, rhombus/diamond) 30
   - Concave vs. convex 31

   Polyhedrons/polyhedra and other three-dimensional shapes
   - Tetrahedron, pentahedron, hexahedron (cube), heptahedron, octahedron, decahedron, dodecahedron, icosahedron (Vocabulary: faces) 32
   - Parallelepiped, diagonal of a hexahedron 36
   - Solid figures that are not classified as polyhedrons (Cylinder, cone, sphere) 37
   - Prisms vs. pyramids 38
   - Right vs. oblique 39
   - Euler's Formula; number of vertices, edges, and faces of solid figures 40

2. **CHAPTER 2 (ANGLES AND DIAGONALS)** **43**
   - Angles (Right, straight, zero angle, acute, obtuse)
     (Vocabulary: vertex, rays) 44
   - Drawing and measuring angles 46
   - Complementary and supplementary angles, linear pair 48
   - Exterior and interior angles 52
   - Angle bisectors, perpendicular bisectors, midpoint 53
   - Constructing congruent line segments, constructing perpendicular lines, constructing triangles with sides that are congruent to given line segments, constructing congruent angles, constructing parallel lines 56
   - Using construction to divide a line segment into three congruent sections 64
   - Finding the measure of bisected angles 66
   - Transversal, alternate interior angles, alternate exterior angles, corresponding angles, vertical angles, adjacent angles 69
   - Parallelograms
     (opposite angles and sides, consecutive angles and sides) 77
   - Diagonals of parallelograms, rhombuses, rectangles, squares, and other convex polygons 78
   - Length of a diagonal (Rectangular solids) 82
   - Length of a diagonal (Cube) 83
   - The sum of a convex polygon's interior angles 84

3. **CHAPTER 3 (ANGLES AND TRIANGLES)** **87**
   - Central angles of regular polygons (Vocabulary: vertex, base angles) 89
   - Triangles (Right, acute, obtuse, oblique triangles) 90
   - Triangles (Equilateral and equiangular, isosceles, scalene) 91
   - The sum of a triangle's interior angles 92
   - Base, legs, hypotenuse, vertex and base angles 98
   - Triangle inequality rule 100
   - Pythagorean Theorem (Theorems and postulates/axioms),
     Pythagorean triple 102
   - Two special right triangles
     (45°-45° right triangle, 30°-60° right triangle) 115
   - Finding a triangle's shortest, longest, and second-longest side 121
   - Similar and congruent 122
   - Triangle congruence properties (SSS, SAS, SAAA, HL) 124
   - Triangle similarity properties (SSS, AA, SAS) 126
   - Similar polygons 127
   - Scale factors 128
   - Using proportions to find missing side lengths of similar triangles 129
   - More similar triangles 133
   - Median 138
   - Perimeter and semiperimeter 139

4. **CHAPTER 4 (CIRCLES AND SPHERES)** **141**
   - Circles (Chord, diameter, radius) 142
   - The measure of a circle and its arcs
     (Minor arcs, major arcs, semicircles) 144
   - Word problems involving circles 145
   - Tangent and secant lines, inscribed and circumscribed triangles 149
   - Word problems involving quadrilaterals inscribed in circles 150
   - Word problems involving circumscribed triangles 152
   - The measurement of an angle formed when two lines (secants or
     tangents) intersect a circle and intersect each other <u>outside the circle</u>,
     <u>inside a circle</u>, or <u>on the circle</u> 155
   - Central and inscribed angles 158
   - More problems involving tangent and secant lines 159
   - Finding the length of an arc 162
   - Pi and circumference 169
   - Equations with pi, in terms of pi 170
   - Area of a circle, surface area of a sphere 172
   - Area of a sector (pie slice) 177
   - More about tangent lines 180

5. **CHAPTER 5 (AREA AND SURFACE AREA)** **182**
   - Definition of area, surface area, and volume 183
   - Area formulas 184
   - Area of a square, rectangle, or parallelogram (Vocabulary: altitude) 185
   - Area word problems 186
   - Area of a rhombus 188
   - Finding a rhombus' height 189
   - Area of a triangle 190
   - Finding a triangle's height 195
   - Area of a trapezoid 197
   - Finding a trapezoid's height 198
   - Area of a regular polygon (Terms: radius, apothem) 199
   - Area of a square, surface area of a cube 200
   - Word problems involving area and perimeter of squares and
     Rectangles 201
   - Lateral surface area 210
   - Surface area (Cylinder) 211
   - Surface area (Right circular cone) 212
   - Surface area (Regular pyramid) 213
   - Slant height and surface area problems 214
   - Surface area (Prism) 219
   - Platonic solids (Surface area of regular polyhedra—tetrahedron,
     hexahedron, octahedron, dodecahedron, icosahedron) 220

6. **CHAPTER 6 (VOLUME, TRIGONOMETRY, SETS, AND MORE)** 221
   - Volume
     - Volume formulas..222
     - Volume of a rectangular prism (cuboid) 224
     - Volume of a prism or cylinder 226
     - Volume word problems..227
     - Volume of a pyramid or cone 229
     - Volume of a cube..231
     - Volume of a sphere...232
   - Length measurements (Millimeters, centimeters, inches, feet, yards, miles)..234
   - Trigonometry (cos, sin, tan)...236
   - Sets
     - List method and set-builder notation 240
     - Set notation..241
     - Equal sets, equivalent sets, elements, universal set, union, intersection, complement 242
     - Subset, proper subset, null set..243
     - Venn diagram 245
   - Equator, latitude, longitude, prime meridian, and International Date Line..246
   - Statements..
     - Conditional and biconditional 247
     - Converse, inverse, contrapositive 248
     - Symbols (Not/negation, there exists, all).................................249
   - Mathematical Reasoning
     - Inductive reasoning, deductive reasoning, appeal to tendency (Terms: premises, conclusion) 250
     - Valid and invalid arguments..251
     - Sound and unsound arguments, counterexamples 252

REVIEWING HOW TO PLOT POINTS ON A COORDINATE PLANE 253

Class Supplies List for Students

1. Three-ring binder
2. Notebook paper
3. Pencils and erasers
4. Pencil sharpener
5. Zipper pencil bag
6. Ruler
7. Math compass
8. Protractor
9. Globe
10. Calculator (Recommended: CASIO fx-9750GIII)
11. Textbook--
    Geometry Textbook for Middle and High School Students, by 70 Times 7 Math
12. Classwork/homework book--
    70 Times 7 Math: Geometry Classwork/Homework for Middle and High School Students
13. Access to computer-based tests (See pages 14-15 for details.)

The classwork/homework book to supplement this textbook can be purchased by visiting the "Math Materials" page of the Habakkuk Educational Materials website at the address below. The answer key for teachers can also be purchased from the website.

https://www.habakkuk.net/

70 Times 7 Math: Geometry Classwork/Homework (for middle and high school students)

70 TIMES 7 MATH (GEOMETRY TEXTBOOK FOR MIDDLE AND HIGH SCHOOL STUDENTS)

The 70 Times 7 Math curriculum begins teaching geometry concepts to children as young as preschool. This textbook makes allowances for it to be studied thoroughly by middle and high school students. If you are following the suggested schedule for teaching the 70 Times 7 Math curriculum (available on the "70 Times 7 Math Curriculum" page of the Habakkuk Educational Materials website), this book would first be studied by eighth-grade students and then restudied by tenth graders. Thus, most, if not all, of the book would be review for tenth-grade students in this program, maximizing their potential for success.

You can review the table of contents to note the wide variety of topics students will learn about in this book. It essentially covers the same material you would expect to find in any other geometry textbook for middle and high school students but in fewer pages and with an approach that is easier to understand. Please do note that while only one textbook is needed, the homework books and the computer-based tests and practice tests that supplement the program are sold separately. (The chart on the following page specifies the materials that are available for this subject.) The 70 Times 7 Math program eliminates the hassle and the cost of purchasing different textbooks at every grade level without compromising what students are required to learn in each grade. Moreover, this curriculum allows for constant review, both from day-to-day and from year-to-year, which is meant to ensure that EVERY student has ample time to gain proficiency in math and to meet state standards. An overview of how the geometry material is introduced is outlined in the pages that follow.

"70 TIMES 7 MATH CURRICULUM" THAT CAN ALSO BE USED TO SUPPLEMENT ANY OTHER SCHOOL-BASED CURRICULUM: While this textbook is complete in that it provides everything necessary to teach geometry to middle and high school students, some of Habakkuk's resources can also be used to supplement any other school-based curriculum. For example, any geometry teacher could help his/her students to meet state standards simply by working through the pages of *70 Times 7 Math: Electronic Textbook for Teachers (Geometry for Middle and High School Students)* on an interactive whiteboard a couple of days a week. (As will be explained in pages to come, this is not a typical textbook, as it is made up of math activities and magnified problems, which can be solved on an interactive whiteboard when using the electronic version for teachers. It is referred to as a textbook because it thoroughly and in a straight-to-the-point manner teaches students what they are required to learn in a middle and high school geometry course.) Another option would be for the teacher to skip directly to a page on the whiteboard that teaches how to solve problems they are currently studying in their own textbooks.

The chart on the next page outlines the materials available for a geometry class, and the chart on page 13 provides schedules that can be followed to help students proceed through the material. The geometry schedule is divided into 5-week sections and specifies what **textbook pages** (or **electronic textbook pages**) need to be covered during that timeframe, the four **classwork assignments** that should be completed (one each week), the **test** that should be administered at the end of the five-week timeframe, and assignments to review. Student textbooks can be purchased in electronic, paperback, or hardback form. The electronic textbooks come with tutorial videos. Please note that answers to problems to be solved on the screen are not included in the electronic textbook for teachers but are provided in the student editions if you want to use them as a reference. Resources listed in the columns of the chart are sold separately from the textbook and can be purchased by visiting the Habakkuk Educational Materials website at https://www.habakkuk.net/.

Subject/Grades	Student Textbook (electronic, paperback, or hardback)	Electronic textbook for teachers	Classwork/ Homework Book	Answer Keys	Computer-Based Tests
Geometry for Middle and High School Students	*Geometry Textbook for Middle and High School Students*, by 70 Times 7 Math	*70 Times 7 Math: Electronic Textbook for Teachers (Geometry for Middle and High School Students)* This is used for direct instruction purposes on a computer screen or interactive whiteboard.	*70 Times 7 Math: Geometry Classwork/ Homework (for middle and high school students)* **Note:** Online practice tests are also available. (See pages 14-15 to learn more.)	*70 Times 7 Math: Answer Keys to Geometry Classwork/ Homework (for middle and high school students)*	The computer-based chapter tests are available through an LMS. (See pages 14-15 to learn more.)

70 TIMES 7 MATH (GEOMETRY TEXTBOOK FOR MIDDLE AND HIGH SCHOOL STUDENTS)

The textbook is available as an electronic book, a paperback, and as a hardback. If you are following the suggested schedule for teaching the 70 Times 7 Math curriculum (available on the "70 Times 7 Math Curriculum" page of the Habakkuk Educational Materials website), this book would first be studied by eighth-grade students and then restudied by tenth graders. The student editions can be used by both students and teachers, as they provide the answers that have been omitted from the corresponding electronic textbook that teachers use on an interactive whiteboard to guide students through the problems. The electronic textbook for students is identical to the print form but also comes with tutorial videos, which students can click on to watch a teacher as she explains how to solve the problem(s) on the page.

70 TIMES 7 MATH: ELECTRONIC TEXTBOOK FOR TEACHERS (GEOMETRY FOR MIDDLE AND HIGH SCHOOL STUDENTS)

This electronic textbook for teachers is used by the teacher for direct instruction purposes and allows her, amongst other things, to display the pages of the textbook onto an interactive whiteboard while using an electronic pen to solve the magnified problems with her students. The difference between this and the student edition is illustrated below. Essentially, answers have been omitted from the teacher's edition so that the problems can be solved on an interactive whiteboard in the presence of the students. Although the answers are hidden, step-by-step directions for solving the problems are provided. In contrast, the student edition illustrates the step-by-step process for solving the problems and also provides the answers, as can be seen in the illustrations to the right. As textbook pages like those shown in the illustrations to the left are displayed onto an interactive whiteboard, the teacher (or a student called upon by the teacher) would use an electronic pen to guide students through the problems, rewriting the steps and the final answers large enough to be seen by the class. The teacher would then go on to the next page and solve the given problems in like manner. Summarized directions on how to use the *70 Times 7 Math: Electronic Textbook for Teachers* are provided on page 16.

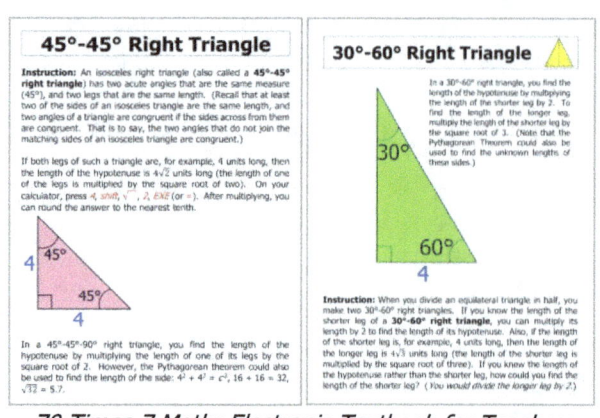

70 Times 7 Math: Electronic Textbook for Teachers (Geometry for Middle and High School Students)

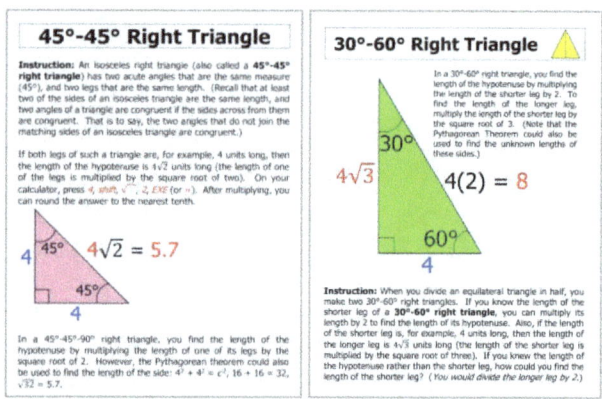

Geometry Textbook for Middle and High School Students, by 70 Times 7 Math

THE DESIGN OF THE TEXTBOOK PAGES: One of the most unique things about this curriculum is the design of the textbook pages. It is referred to as a textbook because it thoroughly and in a straight-to-the-point manner teaches students what they need to know to become proficient in geometry. However, many of the pages (which can be displayed onto an interactive whiteboard) are nothing like a traditional textbook. You will find magnified problems, complete with directions for the teacher, that are ready to be solved by the teacher or a student at the turn of each page. (The answers and the step-by-step process for solving the problems are available in the paperback, hardback, and eBook versions of the student edition.) The design provides an organized and time-efficient way of introducing new material and of reviewing what has already been learned. There are colorful illustrations to help students learn about angles, triangles and other polygons, polyhedrons, circles, spheres, area, surface area, volume, trigonometry, and much more.

HOW TO PROCEED THROUGH THE MATERIAL: As seen on the schedule on page 13, the classwork assignments and tests are scheduled periodically throughout the year. There are a total of six tests, one for each chapter in the book. Let's say that a chapter has a total of 25 pages with general information students need to know and problems they need to know how to solve for a test to be administered five weeks later. The first school day after completing their last test, the teacher would attempt to cover all 25 pages (or as many as possible), working the problems with students on the screen. The second day, he/she would again cover these same 25 pages (or would continue where she left off before going back to the beginning of the chapter). This routine would continue for nearly five weeks (the amount of time students are allotted to prepare for tests), but one day would be set aside each week to review information covered on previous tests so as to help students retain what they have already learned. (Again, this is not a traditional textbook. Although it covers everything a student will need to learn to become proficient in middle and high school geometry, many of the pages take only seconds to go over with students.) While students might originally seem overwhelmed by all the material, it is remarkable how quickly they catch on with the daily reviews.

CONSTANT REVIEWING: Because of the emphasis upon review, this curriculum allows for a smoother transition of students from one grade to another and from one school or school district to another. Even if circumstances prevented a child from learning all he/she should have learned the previous semester or year, he would be given plenty of "second chances" as the information continues to be retaught (or reviewed) and practiced consistently throughout different grade levels. Thus, the constant reviewing, both from day-to-day and from year-to-year, is meant to ensure that EVERY student has ample time to gain proficiency in math and to meet state standards.

CHAPTERS OF THE BOOK: The textbook is divided into six chapters. Most, if not all, the material would be review for any tenth grader who started this program by the 8th grade, as this book is first introduced in this grade, and some of its content is even taught throughout the elementary grades in this program. Thus, this program allows for constant review in the different grade levels. Geometry objectives students are required to master in the elementary grades, for example, are covered in the elementary grades and then reviewed in the 8th and 10th grades. Geometry objectives that students are required to master by the eighth grade are first introduced in the eighth grade and then reviewed by tenth-grade students.

TUTORIAL VIDEOS IN THE ELECTRONIC TEXTBOOKS: Students can click on the tutorial videos from the pages of their electronic textbooks to watch a teacher as she explains how to solve a problem on a given page or even to listen to one of the learning songs taught with this curriculum. Since the classwork/homework assignments list the page numbers where the problems are explained, students can get there quickly and may opt to watch the tutorial videos when the teacher is unavailable. This resource not only benefits traditional students who need help with their classwork but can also serve as a math teacher for home-schooled students.

70 TIMES 7 MATH: GEOMETRY CLASSWORK/HOMEWORK

There are four classwork (or homework) assignments to help prepare students for each test. (Computer-based practice tests for each exam are also available by visiting the website of Habakkuk Educational Materials.) Books with these assignments are sold separately. Students should complete one of the assignments weekly and then take the test at the end of the fifth week. For homework the fifth week, they can correct any errors from previous assignments. Each week's classwork will be in the same format as the test. If there are, let's say, 100 problems on an upcoming test, the student will complete all 100 of those types of problems each week for four weeks, although the answers to corresponding problems will differ from one week to the next. Questions included on the classwork and tests are not randomly selected from the chapters in focus. Instead, they are very comprehensive in that every type of problem students learn about in their textbook and complete as a class on the interactive whiteboard will also be solved individually by students on the classwork assignments and corresponding tests. Moreover, each classwork/homework assignment specifies the page numbers in their textbooks where students can turn to for help as needed.

70 TIMES 7 MATH: GEOMETRY TESTS (FOR MIDDLE AND HIGH SCHOOL STUDENTS)

There are six comprehensive computer-based tests specially designed to assess a student's knowledge of the information covered in this book. Since students are tested over every type of problem included in their textbooks, some of these comprehensive tests are somewhat lengthy and may take more than one class period to complete if you have a fixed amount of time for math each day. Please allow every student to finish each test without penalty. Note that tests for students in grades 6th-12th are only available through enrollment in an LMS. (Please see page 14 for enrollment information.) If you prefer, one of the classwork/homework assignments from *70 Times 7 Math: Geometry Classwork/Homework (for middle and high school students)* could instead be used for testing purposes.

70 TIMES 7 MATH: ANSWER KEYS TO GEOMETRY CLASSWORK/HOMEWORK (FOR MIDDLE AND HIGH SCHOOL STUDENTS)

Answer keys for the classwork/homework assignments are sold separately.

GEOMETRY SCHEDULE (Suggested Grades: 8th and 10th)

GEOM (stands for geometry)

Classwork/Homework: There are four classwork/homework assignments to help prepare students for each test. Students should complete one of the assignments weekly for four weeks and then spend the 5th week correcting any errors from them. The test can be administered at the end of the 5th week.

	Textbook/Electronic Textbook chapters to cover	Classwork/ Homework	Online Tests	Computer-Based Reviews
Weeks 1-5	CHAPTER 1	GEOM Classwork, #1A GEOM Classwork, #1B GEOM Classwork, #1C GEOM Classwork, #1D	Geometry Test, #1 (Administer at the end of the 5th week.)	NOTE: THE COMPUTER-BASED PRACTICE TESTS CAN BE USED FOR THE REVIEWS NOTED BELOW.
Weeks 6-10	CHAPTER 2	GEOM Classwork, #2A GEOM Classwork, #2B GEOM Classwork, #2C GEOM Classwork, #2D	Geometry Test, #2 (Administer at the end of the 10th week.)	Review (#1, Geometry Practice Test)
Weeks 11-15	CHAPTER 3	GEOM Classwork, #3A GEOM Classwork, #3B GEOM Classwork, #3C GEOM Classwork, #3D	Geometry Test, #3 (Administer at the end of the 15th week.)	Review (Geometry Practice Tests: #1 and #2)
Weeks 16-20	CHAPTER 4	GEOM Classwork, #4A GEOM Classwork, #4B GEOM Classwork, #4C GEOM Classwork, #4D	Geometry Test, #4 (Administer at the end of the 20th week.)	Review (Geometry Practice Tests: #1, #2, and #3)
Weeks 21-25	CHAPTER 5	GEOM Classwork, #5A GEOM Classwork, #5B GEOM Classwork, #5C GEOM Classwork, #5D	Geometry Test, #5 (Administer at the end of the 25th week.)	Review (Geometry Practice Tests: #1, #2, #3, and #4)
Weeks 26-30	CHAPTER 6	GEOM Classwork, #6A GEOM Classwork, #6B GEOM Classwork, #6C GEOM Classwork, #6D	Geometry Test, #6 (Administer at the end of the 30th week.)	Review (Geometry Practice Tests: #1, #2, #3, #4, and #5)
Weeks 31-32	Review			Review

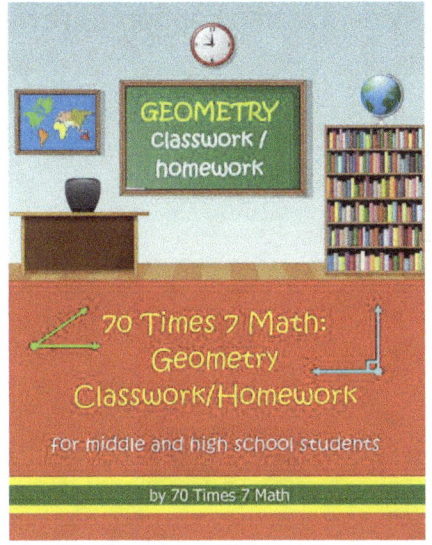

70 Times 7 Math: Geometry Classwork/Homework (for middle and high school students)

COMPUTER-BASED TESTS FOR THE "70 TIMES 7 MATH CURRICULUM"

Overview of the online tests and how to gain access: Computer-based tests specially designed to assess a student's knowledge of the information covered in this book are available on a Learning Management System. To gain access to the tests, please visit the website of Habakkuk Educational Materials at https://www.habakkuk.net/ and click on "Math Materials" from the homepage. There you will find an online request form to fill out and a flyer to download with more information about these online tests. Habakkuk Educational Materials will finalize your enrollment within 48 hours of purchase (please refer to the form for pricing details), and you will have complete access to the online tests during the enrollment period, which includes both practice tests and a regular test for each chapter of the textbook. The four classwork/homework assignments for each chapter of the book are in the same format as the tests and one could be used in place of the computer-based tests if you prefer a paper version. The title of this book is *70 Times 7 Math: Geometry Classwork/Homework (for middle and high school students)*. Please note that questions included on the classwork and tests are not randomly selected from the chapters in focus. Instead, they are very comprehensive in that every type of problem students learn about in their textbook or complete as a class on the interactive whiteboard will also be solved individually by students on the classwork assignments and corresponding tests.

How to use the tests: It is recommended that students originally be allowed to use their books to guide them through the practice tests. When the student feels confident or at the teacher's discretion, there is also a computer-based chapter test they can take to test their comprehension over the book's content. After submitting an answer to a question, the computer will notify test takers if their answer is correct or incorrect. When taking the practice tests, students are often given a second chance to answer the question correctly, and tutorial videos are available to provide them assistance as needed. After entering their answer to the final question of a test or practice test and clicking "Continue," a percentage grade will be available and a "Review Course" option to review any incorrect answers will also be accessible.

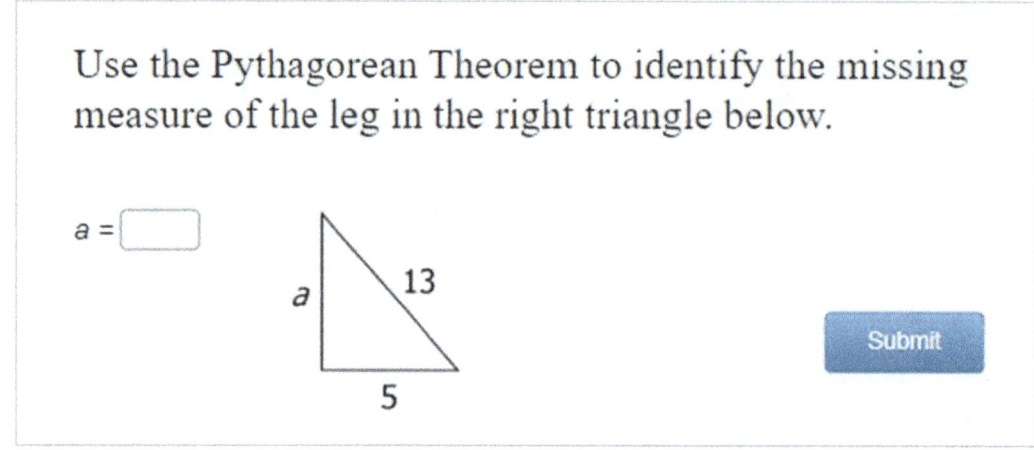

The tests serve two purposes:
(1) to assess a student's mastery of the content in this book; and
(2) to be used for review games.

Using the online practice tests for pretests and for review: The practice tests can be used as **pretests** at the beginning of a school year to determine what students already know, and parents could be given a summary of the results. The practice tests also provide an opportunity for students to independently review material covered on previous tests. Beginning the first week following their first test, students spend once a week at the computers answering questions over material that they have already been tested over so as not to forget it once the new information has been introduced.

Using the online tests for review games: The computer-based tests can be used in correspondence with various board games that have a pathway from start to finish. The review games could be played in math centers, or the entire class could be divided into small groups and a gameboard provided for each set of students. Children playing the game would answer one of the questions, and if the computer confirms that the answer is correct, the student could roll dice or spin a spinner and move his or her playing piece the corresponding number of spaces on the path.

If you visit the "Free Teaching Materials" page of the Habakkuk Educational Materials website, there are free gameboards that can be printed on cardstock. (Click on "Board games and more to complement Habakkuk Educational Materials' Bible, reading, language, math, science, and social studies materials.") The coupons referred to in the directions of one of the gameboards can be printed from the same file, and students who earn a "Good Work Coupon" might be designated a certain amount of time to play math games on the computer or whatever other privilege the teacher might choose.

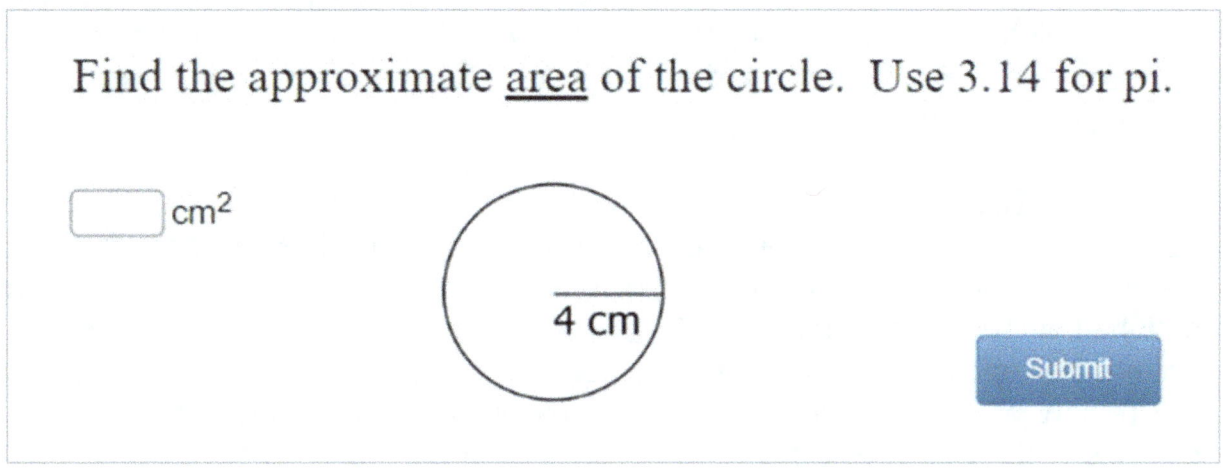

TEACHER DIRECTIONS: Open *70 Times 7 Math: Electronic Textbook for Teachers (Geometry for Middle and High School Students)* on your classroom screen. Proceed through the pages of the chapter you are studying by following the given instructions. There are a total of six tests, one for each chapter in the book. Let's say that a chapter has a total of 25 pages with general information students need to know and problems they need to know how to solve for a test to be administered five weeks later. The first school day after completing their last test, the teacher would attempt to cover all 25 pages (or as many as possible), working the problems with students on the screen. The second day, he/she would again cover these same 25 pages (or would continue where she left off before going back to the beginning of the chapter). This routine would continue for five weeks (the amount of time that geometry students have to prepare for a test), but one day should be allotted each week to review information covered on previous tests so as to help students retain what they have already learned.

If you are a **homeschooling parent** using the paperback or hardback to teach your students, it is recommended that you place a sheet of transparency paper over any pages before writing on them with overhead projector pens so that they can be reused during your daily reviews. Problems that are already solved in their textbooks could also be solved step-by-step on paper or on a mini whiteboard.

ASSIGN HOMEWORK OR CLASSWORK FOR INDIVIDUAL PRACTICE: Students studying from this textbook should also have a copy of *70 Times 7 Math: Geometry Classwork/Homework (for middle and high school students)*. There are four classwork (or homework) assignments to help prepare students for each test. (As discussed on the two previous pages, computer-based practice tests are also available by visiting the website of Habakkuk Educational Materials.) Students should complete one of the assignments weekly and then take the test at the end of the fifth week. For homework the fifth week, they can correct any errors from previous assignments.

REVIEW MATERIAL
   To help students retain what they have already learned, allow one day each week to review pages covered on previous tests. If you are able to review the first half of the first chapter the first week, you might attempt to review the second half of the same chapter the second week. In other words, continue the review where you left off the previous week. Using the computer-based practice tests from previous chapters is also a good source of review.

Chapter 1

Plane and 3-D Geometry

(Suggested Grades: 8th and 10th)

Note to teachers: When studying chapter 1, please also take time to go over page 77 with your class, as it contains information students need to know when completing certain classwork/homework and test problems for this chapter.

Teacher instructions: Using *70 Times 7 Math: Electronic Textbook for Teachers (Geometry for Middle and High School Students),* ask students to identify any missing answers for you to write on the screen. Please note that since the answers are provided in student textbooks, they should have them closed during this time. Student textbooks can also be used as a key for the teacher's benefit.

Instruction: Geometry studies points, lines, and planes and the shapes they create.

point
• *A*

Instruction: This point is named *A*. A capital letter is used to name a point.

line: $\overleftrightarrow{AB}$

Instruction: This is line *AB*. The arrowheads signify that the line continues endlessly in both directions. A line has at least two points. *Collinear* means lying in the same line. Do you see the word *line* in it?

plane: *ABC*

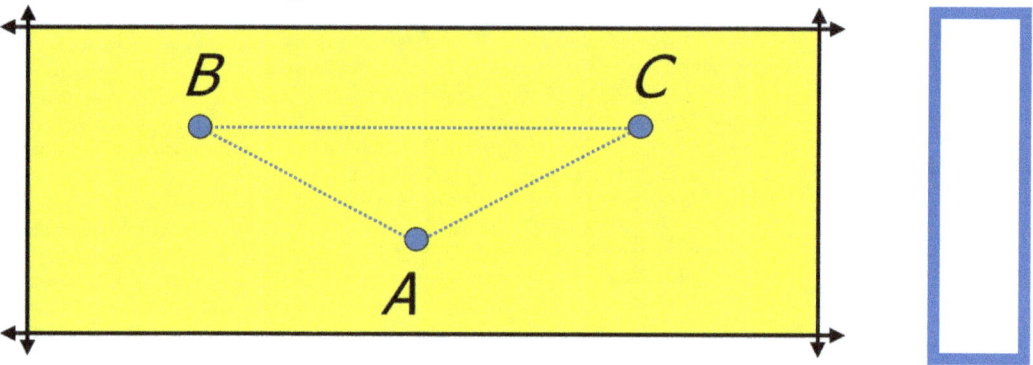

Instruction: A plane is flat and continues endlessly in every direction. Think of the yellow box as being a plane. A unique plane has at least three points that are not all on the same line. Think of the triangle (plane *ABC*) as being a unique plane; two of its points are on the same line and the other is not. The rectangle is a plane with 4 points. (Illustrate the four points, or vertices, of the rectangle.) *Coplanar* means lying in the same plane. You can see that it almost has the word *plane* in it.

space

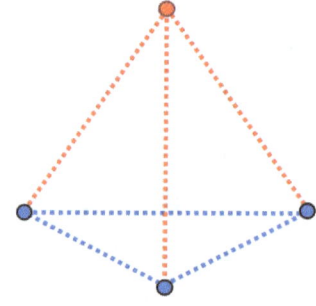

Instruction: Space has at least four points that are not all lying in the same plane. In the tetrahedron to the right, the three points of the blue triangular base are on the same plane, but the fourth point is not.

Lines

intersecting

Instruction: As indicated by the name, intersecting lines intersect. Lines that intersect are coplanar lines, meaning that they lie in the same plane. ***Concurrent*** lines intersect at a single point.

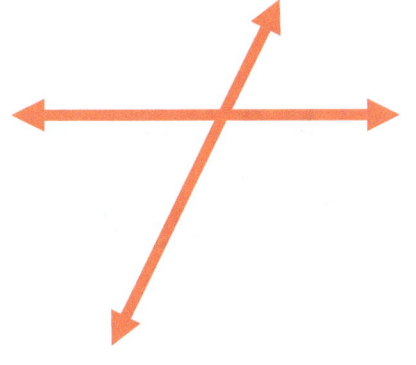

perpendicular

Instruction: Notice from the illustration that perpendicular lines are also intersecting lines, but they intersect to form right angles (the little squares symbolize that they are 90° right angles). The symbol for perpendicular lines is ⊥. The notation $e \perp f$ reads "e is perpendicular to f."

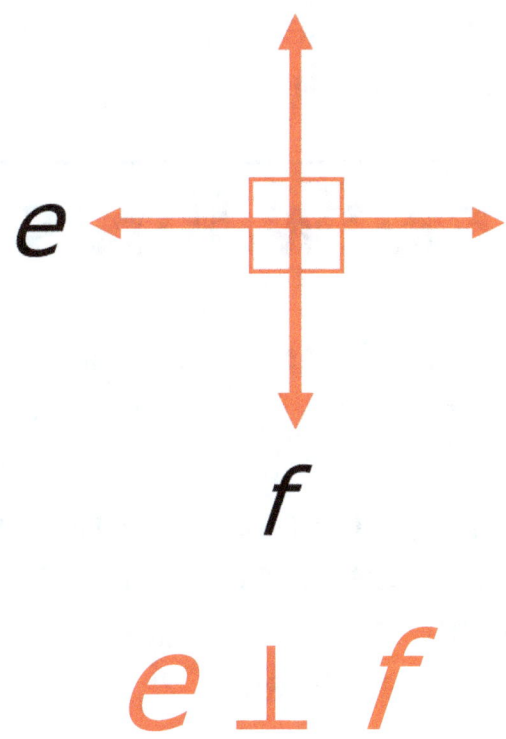

parallel lines

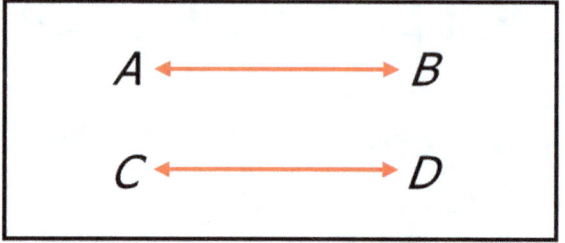

Instruction: The symbol for parallel lines (or parallel planes) is ∥. This reads "line *AB* is parallel to line *CD*." Parallel lines do not intersect, they are the same distance apart, and they are coplanar.

The figure below represents **parallel planes** in space. Like parallel lines, parallel planes do not intersect.

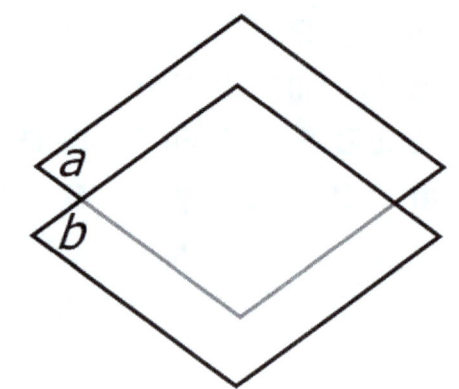

skew lines

Instruction: Skew lines are noncoplanar (meaning that they do not lie in the same plane). The two straight lines cannot be parallel, and they cannot intersect. Lines *s* and *t* are skew lines.

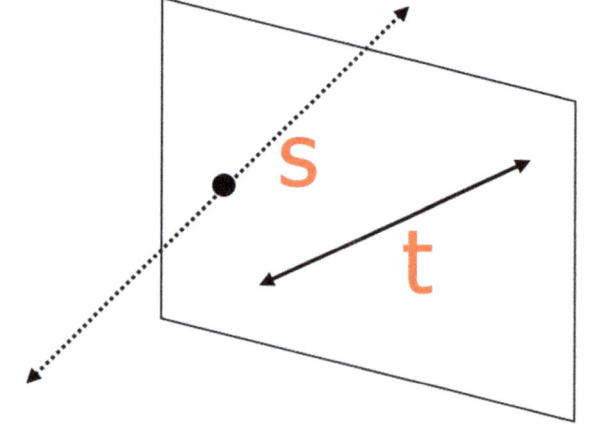

segment AB $\overline{AB}$	**line AC** $\overleftrightarrow{AC}$ or $\overleftrightarrow{CA}$
Instruction: The line segments in the figure below are $\overline{AB}$, $\overline{BC}$, and $\overline{AC}$. Line segment AB could have also been named $\overline{BA}$.	**Instruction:** The line in the figure below can either be named $\overleftrightarrow{AC}$ or $\overleftrightarrow{CA}$.

ray AB
$\overrightarrow{AB}$

Instruction: A ray has just one endpoint and is moving in the same direction. Unlike the line segment and the line above, ray AB in the figure below could not have also been named $\overrightarrow{BA}$, but it could have been named $\overrightarrow{AC}$. Since $\overrightarrow{AB}$ and $\overrightarrow{AC}$ have the same endpoint (A) and are moving in the same direction (to the right), they are the same ray. In the notation, the endpoint is listed before the other letter, and the ray always points to the right, regardless of its direction in the line.

The rays in this figure are $\overrightarrow{AB}$ (or $\overrightarrow{AC}$), $\overrightarrow{CB}$ (or $\overrightarrow{CA}$), and the opposite rays $\overrightarrow{BA}$ and $\overrightarrow{BC}$. Notice that there are a total of four rays in this example. One has **A** as the endpoint and points to the right. A second ray has **C** as the endpoint and points to the left. Two more rays have **B** as the endpoint (one points to the left and the other to the right).

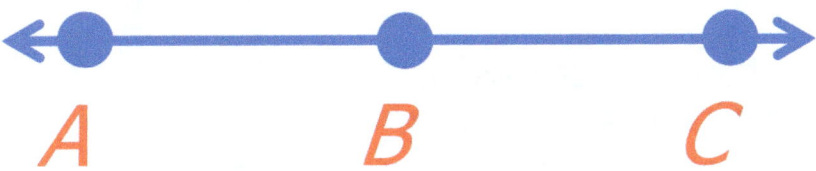

Instruction: A curve can cross itself, but the only way a *simple curve* can intersect itself is when the starting and ending points touch, as in a simple closed curve. Think of it this way, "It's so simple that it can't even cross itself."

curve

closed curve

This shape is also referred to as a "**complex polygon**." A polygon that is complex can cross over itself.

simple curve

simple closed curve

This shape is also called a "**simple polygon**." If a polygon is simple, it cannot cross over itself.

polygonal curve

Instruction: Since this shape is not closed, it is not a polygon. However, it is a polygonal curve because it is made up of line segments.

Polygons

Instruction: A polygon has three characteristics. It is flat (meaning that it is two-dimensional), it is formed <u>entirely</u> of straight lines, and it is closed—there are no open sides. A 3-dimentional shape is not a polygon because it is not flat, and a circle is not a polygon because it has no straight lines. Also, just like a "simple curve" cannot cross itself, a "simple polygon" cannot cross itself either. The polygons shown below are simple polygons. (Note that the two dimensions of a flat shape are length and height, and the three dimensions of a 3-D figure are length, height, and depth.)

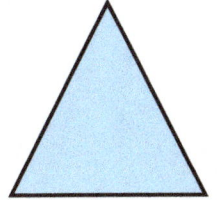 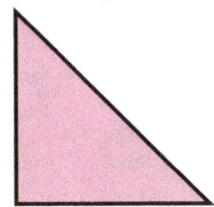  

Regular Polygons

Instruction: A polygon is only regular if it is both **equilateral** and **equiangular**. In other words, its sides are all the same length (equilateral), and its angles are all the same measure (equiangular). If all the sides of a polygon are equal, then its angles must also be equal. Note that if all the sides of a polygon are not equal or if all its angles are not the same measure, then it is an ***irregular polygon***.

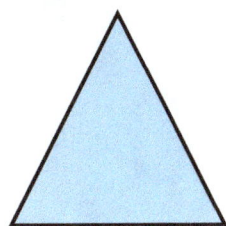 

Instructions: Before viewing the next four pages, click the ear icon to hear "The Polygon Song" sung to the tune of "This Old Man Came Rolling Home." (Paperback readers can hear the tune by visiting the "Math Songs" page of the Habakkuk Educational Materials website at https://www.habakkuk.net/.)

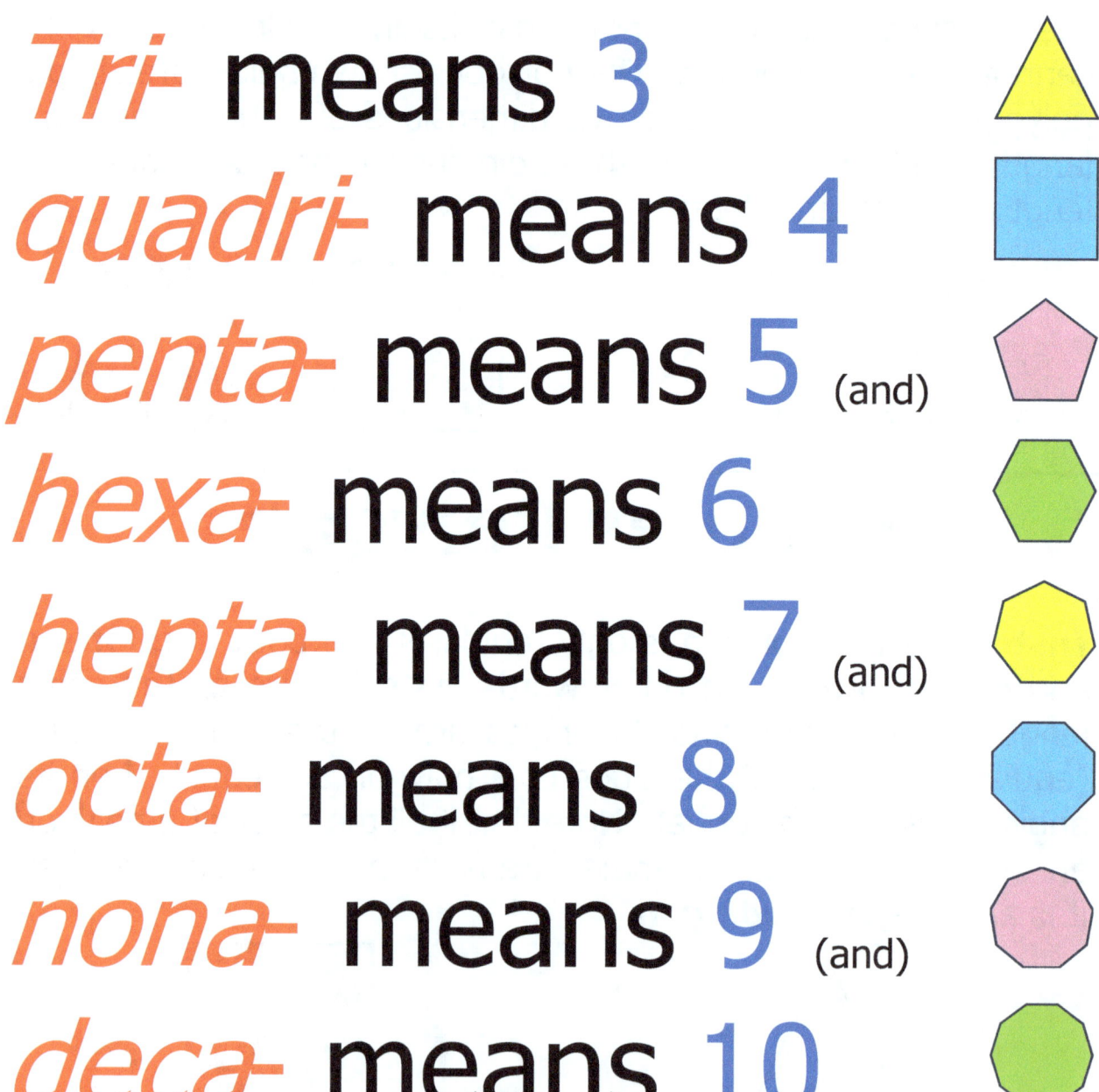

Tri- means 3

quadri- means 4

penta- means 5 (and)

hexa- means 6

hepta- means 7 (and)

octa- means 8

nona- means 9 (and)

deca- means 10

Polygons
(number of sides and angles)

Instruction: You can use the illustrations in this box to review vertices and angles.

A triangle has 3 vertices, one on each corner.

A square has 4 right angles.

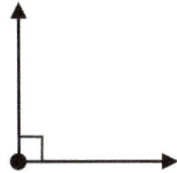

triangle

3

How many <u>sides</u> does a triangle have?
How many <u>angles</u>...?
How many <u>vertices</u>...?

(The answer to all three questions is three.)

quadrilateral

4

How many <u>sides</u> does a quadrilateral have?
How many <u>angles</u>...?
How many <u>vertices</u>...?

(The answer to all three questions is four.)

Examples of quadrilaterals include squares, rectangles, trapezoids, and rhombuses.

Polygons
(number of sides and angles)

pentagon

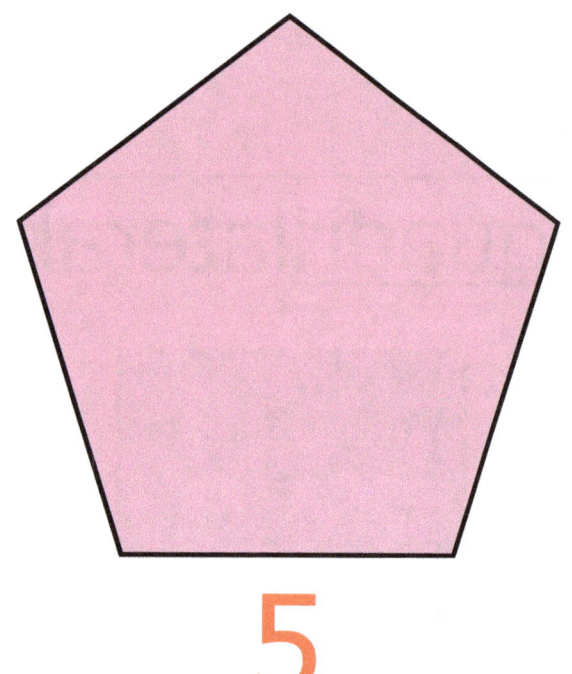

5

How many <u>sides</u> does a pentagon have?
How many <u>angles</u>...?
How many <u>vertices</u>...?

(The answer to all three questions is five.)

hexagon

6

How many <u>sides</u> does a hexagon have?
How many <u>angles</u>...?
How many <u>vertices</u>...?

(The answer to all three questions is six.)

Polygons
(number of sides and angles)

heptagon

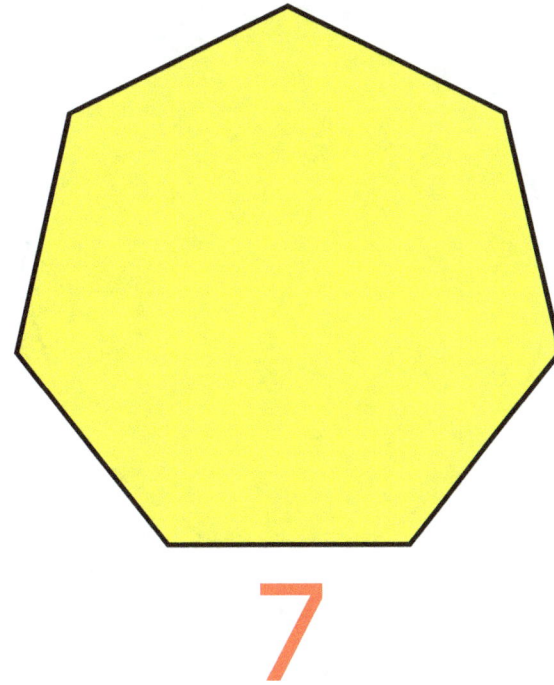

7

How many <u>sides</u> does a heptagon have?
How many <u>angles</u>…?
How many <u>vertices</u>…?

(The answer to all three questions is seven.)

octagon

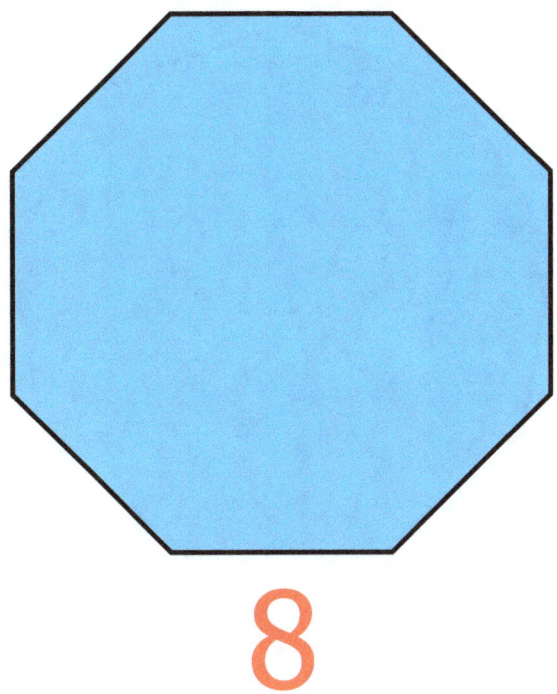

8

How many <u>sides</u> does an octagon have?
How many <u>angles</u>…?
How many <u>vertices</u>…?

(The answer to all three questions is eight.)

Polygons
(number of sides and angles)

nonagon

9

How many <u>sides</u> does a nonagon have?
How many <u>angles</u>…?
How many <u>vertices</u>…?

(The answer to all three questions is nine.)

decagon

10

How many <u>sides</u> does a decagon have?
How many <u>angles</u>…?
How many <u>vertices</u>…?

(The answer to all three questions is 10.)

Polygons
(number of sides and angles)

hendecagon

dodecagon

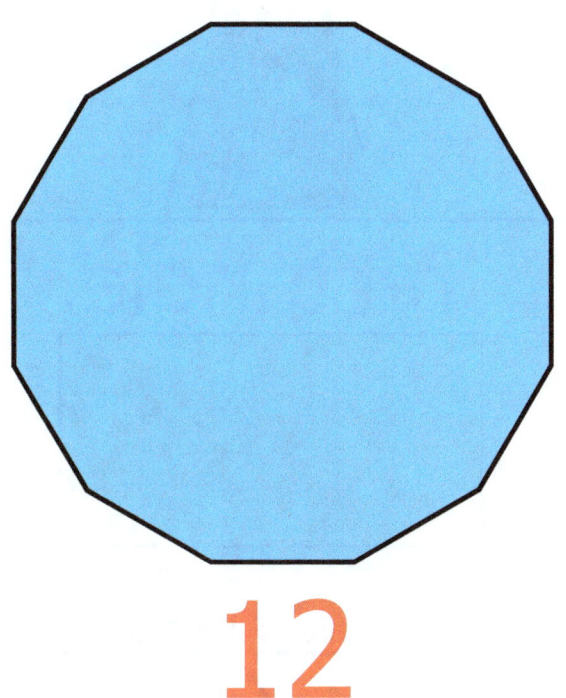

11

12

How many <u>sides</u> does a hendecagon have?
How many <u>angles</u>...?
How many <u>vertices</u>...?

(The answer to all three questions is 11.)

How many <u>sides</u> does a dodecagon have?
How many <u>angles</u>...?
How many <u>vertices</u>...?

(The answer to all three questions is 12.)

Instruction: Explain to students that an **n-gon** is a polygon whose number of sides is not known. A triangle can be referred to as a 3-gon. Can you tell me why? (Answer: It's a polygon with three sides.)

Quadrilaterals 360°

Instruction: All the shapes on this page are quadrilaterals because they have exactly four sides. If you add together what all four interior angles of a <u>convex</u> quadrilateral measure, the sum will always be 360°. (A circle also measures 360°.) You will learn more about this later in the book.

trapezoid

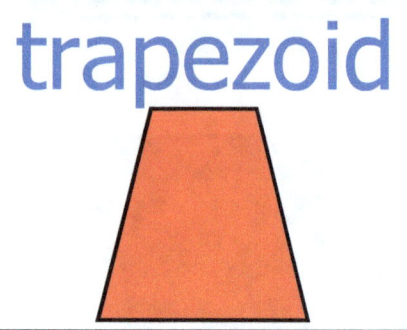

A trapezoid has one pair of parallel opposite sides, called *bases*. In the trapezoid to the left, it is the top and bottom lines that are parallel. Although they are not the same length, they are the same distance apart.

rectangle   square

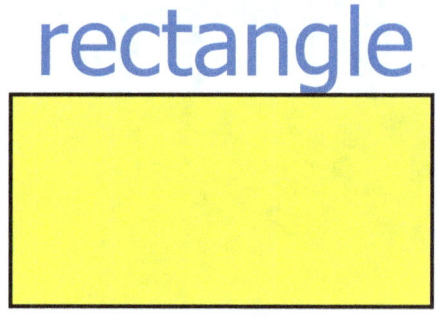 

A square is a rectangle with all four sides the same length.

parallelogram   rhombus
(diamond)

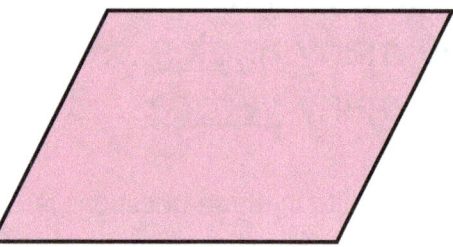 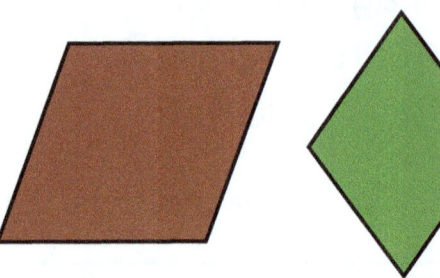

A rhombus (or diamond) is a parallelogram with all four sides the same length. The plural of rhombus is *rhombuses* (or *rhombi*).

concave	**convex**
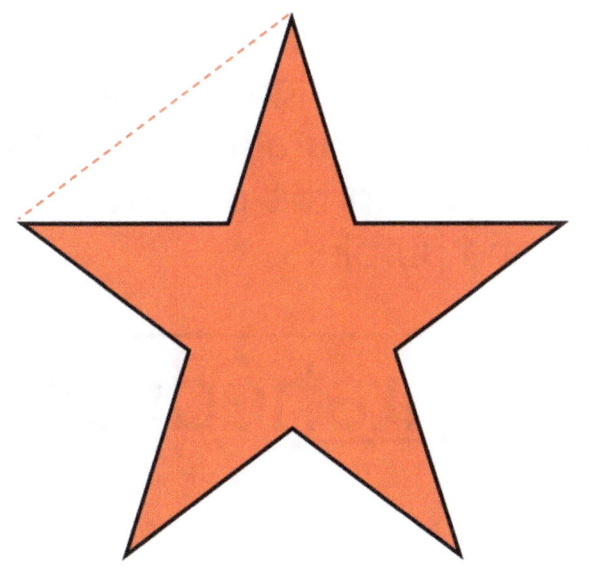	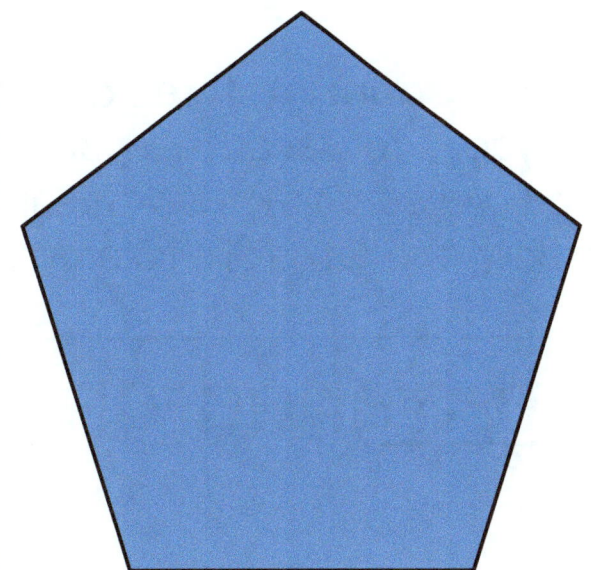
Teacher instructions: Show students that when you draw a line from one point on the star to a point next to it, the line goes outside the shape. Then say, "You can tell that the star is concave because when you draw a line from one point on the star to a point next to it, the line goes outside the shape. A con<u>cave</u> shape looks like it is 'caved' in."	**Teacher instructions:** Show students that a line does not go outside the shape when you draw a line from one of the five points of the pentagon to a point next to it. Then say, "You can tell that the pentagon is convex because when you draw a line from one point on the shape to a point next to it, the line does not go outside the shape."

Polyhedron (number of faces)

Instruction: A three-dimensional (3-D) figure that has polygons for faces is a polyhedron. (The plural is *polyhedra* or *polyhedrons*.) *Poly-* means "many." In the pentahedron below, the faces are colored blue and purple.

tetrahedron

pentahedron

4

5

How many <u>faces</u> does a tetrahedron have?

How many <u>faces</u> does a pentahedron have?

A *tetrahedron* has four faces, and a *pentahedron* has five.

Polyhedron
(number of faces)

Teacher instructions: Help students to understand the concept of faces by letting them count the six faces of a Rubik's Cube or something similar.

hexahedron (cube)

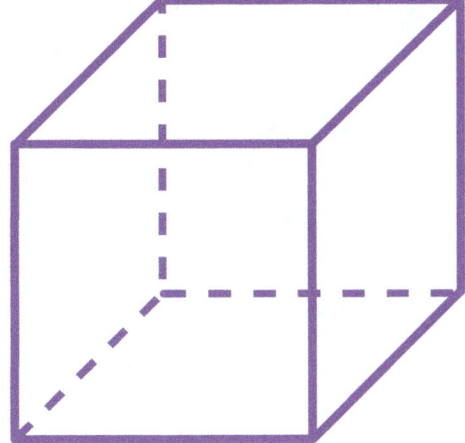

6

How many <u>faces</u> does a hexahedron have?

heptahedron

7

How many <u>faces</u> does a heptahedron have?

A *hexahedron* has six faces, and a *heptahedron* has seven.

Polyhedron (number of faces)

octahedron | decahedron

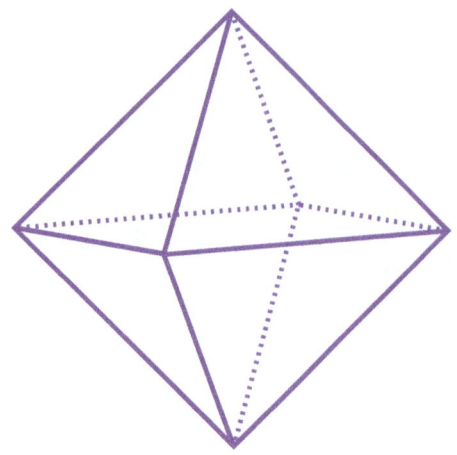 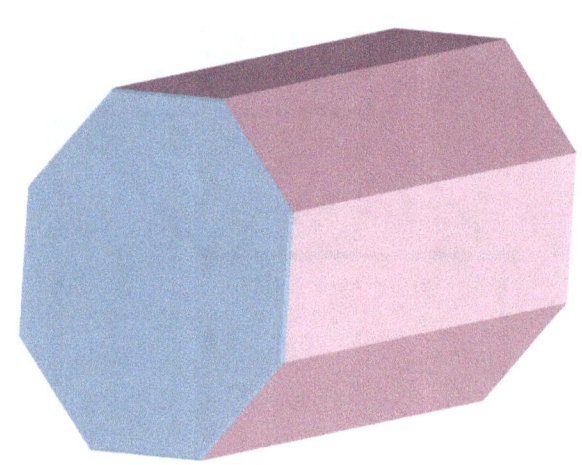

8 | 10

How many faces does an octahedron have? _____ | How many faces does a decahedron have?

An *octahedron* has eight faces, and a *decahedron* has ten.

Polyhedron
(number of faces)

dodecahedron | icosahedron

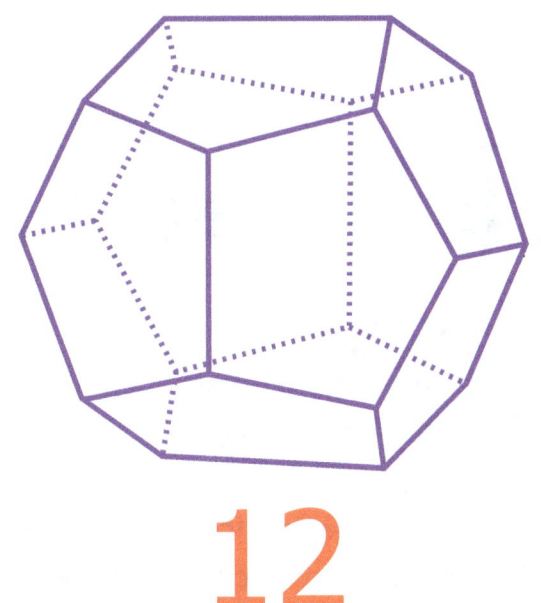 

12 | **20**

How many <u>faces</u> does a dodecahedron have? | How many <u>faces</u> does an icosahedron have?

A *dodecahedron* has 12 faces, and an *icosahedron* has 20.

Parallelepiped

Instruction: In a parallelepiped, each of the six faces of the hexahedron is a parallelogram, and the opposite edges are both parallel and congruent. An example of opposite edges in the parallelepiped below are marked with *x*. Can you identify others?

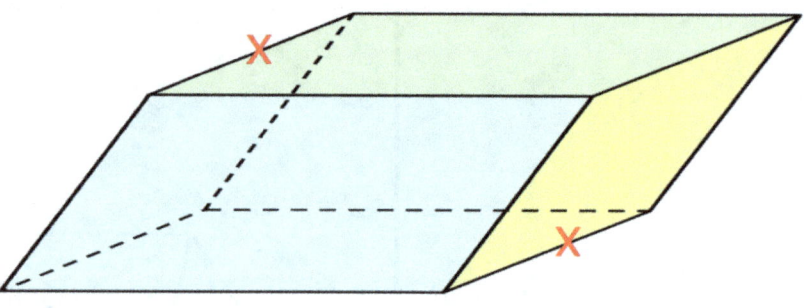

Diagonal of a Hexahedron

A diagonal of a hexahedron (a 3-dimentional figure with 6 faces) is a segment that connects a vertex of one face with an opposite vertex of a different face. Although the diagonals of a parallelepiped are not equal, the four diagonals of a rectangular solid (a solid with rectangular faces like that shown below) are of equal length. Two of its diagonals have been illustrated with red dotted lines. Notice that they cut each other in half. Can you find other diagonals?

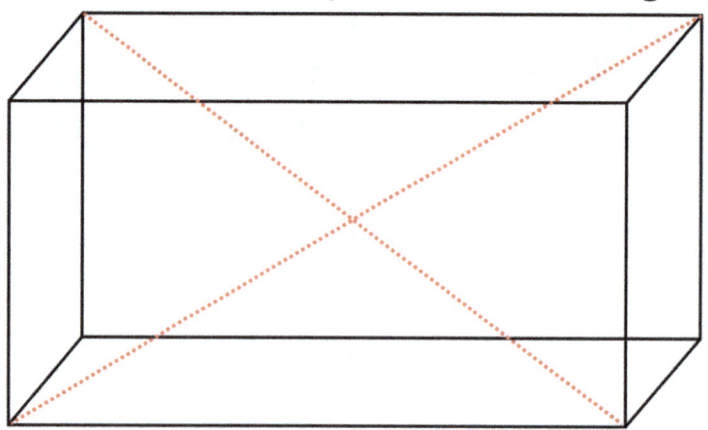

Solid Figures that Are Not Classified as Polyhedrons

Instruction: A geometric solid (or three-dimensional figure) is not a flat shape. Even though the images on this page are solid figures, they are not polyhedrons like the other 3-D figures we just learned about. Do you know why? (The cone, cylinder, and sphere are not polyhedrons since they do not have polygons for faces.)

Identify each solid figure.

cylinder

A cylinder is shaped like a can.

cone

A cone is shaped like an ice-cream cone.

sphere

A sphere is shaped like a ball.

pentagonal prism	pentagonal pyramid
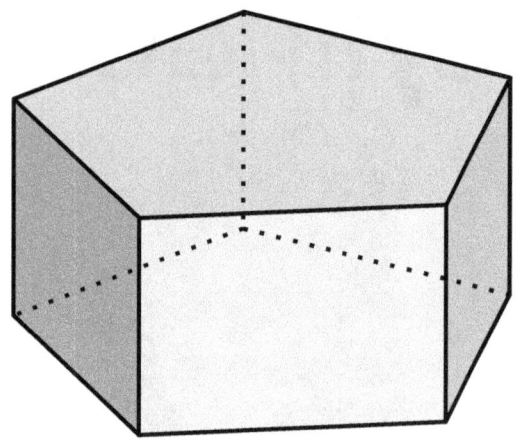	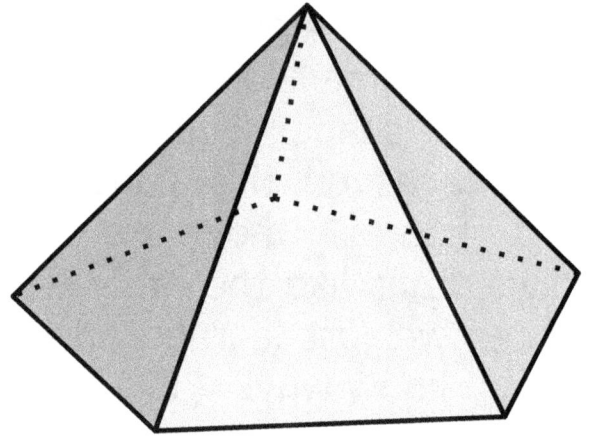
Instruction: A prism has two congruent, polygonal faces, called bases, and all the rest of its faces are quadrilaterals. This is a pentagonal prism because the two congruent faces are pentagons. The height (H) of any prism is the perpendicular distance between both bases.	**Instruction:** The difference between a pyramid and a prism is that a pyramid has just one polygonal base, and the rest of its faces are triangles. We know that this figure is a pentagonal pyramid because its base is a pentagon. Both prisms and pyramids have polygonal bases. **Note:** Since a polygon is a closed, flat shape formed entirely by line segments, a cylinder is not a prism because it has circles for bases.

Right Cylinder

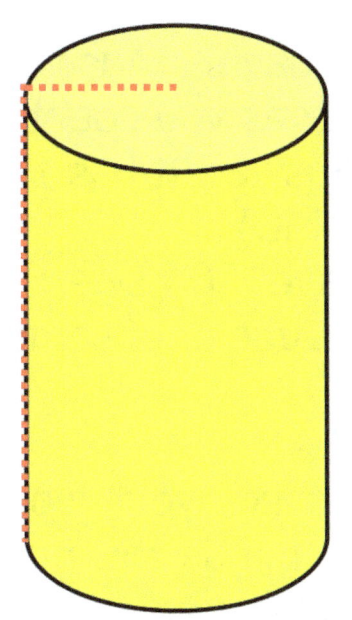

Instruction: Notice from the three-dimensional figure to the left that the segments that form the lateral surface of a ***right*** <u>cylinder (or prism)</u> are at right angles to the base (see a right angle on the first page of the next chapter).

Also, the vertex of a ***right*** <u>cone (or pyramid)</u> lines up with the middle of the base; in an oblique cone or pyramid, it does not.

right cone

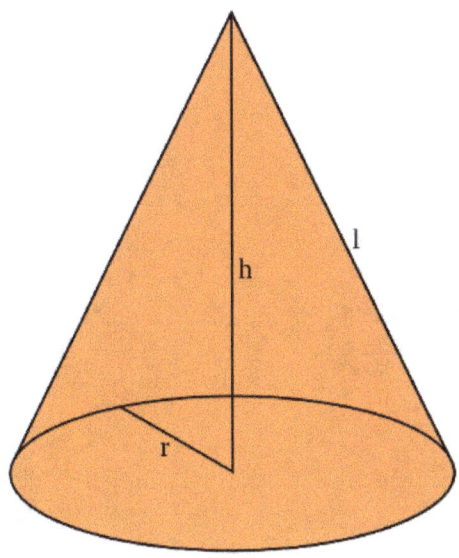

oblique cone

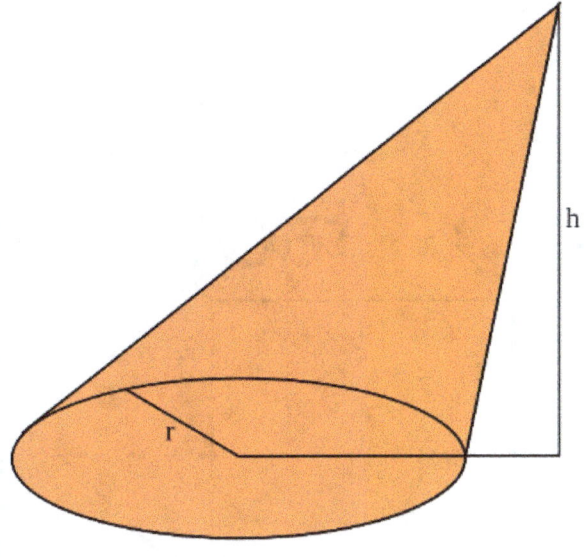

Euler's Formula
$V - E + F = 2$

1. How many **vertices** does this cube have? (Answer: It has 8 vertices. Put a dot on each vertex as you count it.)
2. How many **edges** does it have? (Answer: It has 12 edges. Put an *x* on each edge as you count it.)
3. How many **faces** does the cube have? (Answer: It has 6 faces. Count the faces of a Rubik's Cube or of some similar cube.)

You can use **Euler's formula** to check your work. The formula is the number of vertices, minus the number of edges, plus the number of faces equals two.

$8 - 12 + 6 = 2$

This formula works with convex polyhedrons. It was named after a Swiss mathematician and physicist named Leonhard Euler.

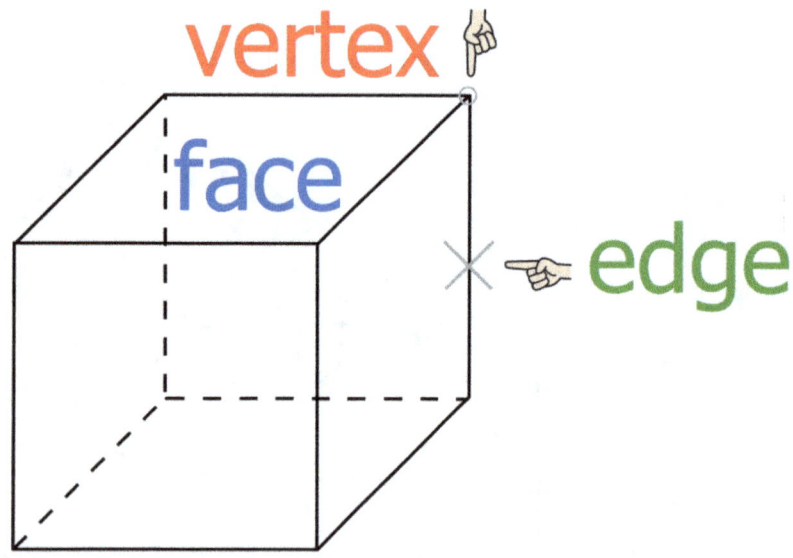

1. Use Euler's formula to find the number of <u>vertices</u> of an octahedron. An octahedron is a polyhedron with 12 edges.
   6
   (**Solution:** __ − 12 + 8 = 2; 2 − 8 = ⁻6, ⁻6 + 12 = 6)

2. Use Euler's formula to find the number of <u>faces</u> of a polyhedron with 4 vertices and 6 edges. Then identify the polyhedron.
   4, tetrahedron
   (**Solution:** 4 − 6 + __ = 2; 2 + 6 = 8, 8 − 4 = 4)
   Note: Since 4 is positive in *4 − 6 + __ = 2*, it is subtracted from the 8.

3. Use Euler's formula to find the number of <u>edges</u> of a hexahedron. A hexahedron is a polyhedron with 8 vertices.
   12
   (**Solution:** 8 − __ + 6 = 2; 2 − 6 = ⁻4, ⁻4 − 8 = ⁻12)
   The answer is 12, not ⁻12, because the subtraction sign will already be in front of where the 12 would be placed in the equation. Besides this, it clearly could not have a negative number of edges.

Notes

Chapter 2

Angles and Diagonals

(Suggested Grades: 8th and 10th)

Note to teachers: When studying chapter 2, please also take time to review page 36 with your class, as it contains information students need to know when completing certain classwork/homework and test problems for this chapter.

Teacher instructions: Using *70 Times 7 Math: Electronic Textbook for Teachers (Geometry for Middle and High School Students),* ask students to identify any missing answers for you to write on the screen. Please note that since the answers are provided in student textbooks, they should have them closed during this time. Student textbooks can also be used as a key for the teacher's benefit.

right angle
90°

Instruction: A right angle measures 90°. The **vertex** of the angle to the right is B. The **rays** are $\overrightarrow{BA}$ and $\overrightarrow{BC}$.

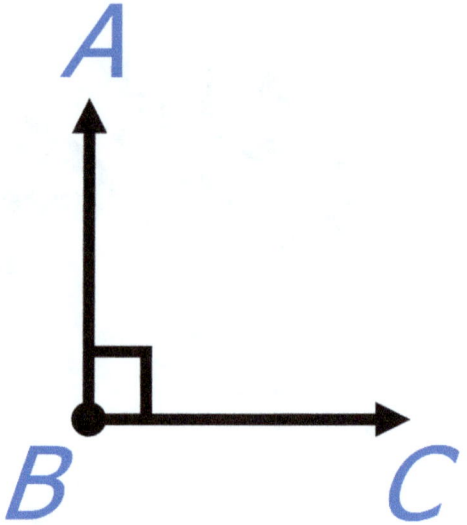

straight angle
180°

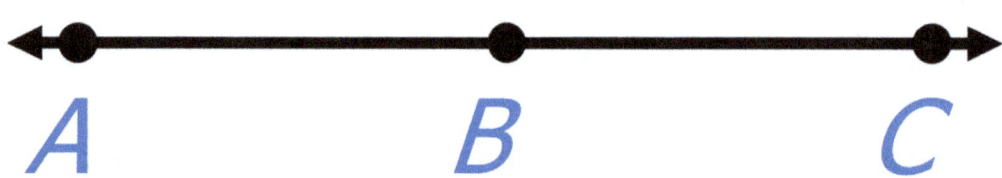

Instruction: A straight angle measures 180°. In a straight angle, the two arrows are pointing in opposite directions.

zero angle
0°

Instruction: Notice that in a zero angle, both arrows are pointing in the same direction.

acute angle

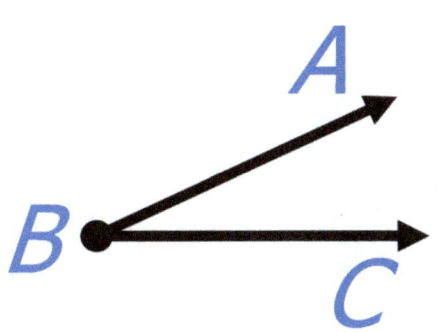

An acute angle measures...

**less than 90°
& more than 0°.**

Teachers call it "<u>a</u> <u>cute</u> little angle." This angle can be named ∠B, ∠ABC, or ∠CBA. Notice that the vertex B is listed by itself or in the middle.

$$\angle B \qquad \angle ABC \qquad \angle CBA$$

---

obtuse angle

An obtuse angle measures...

**more than 90°
& less than
180°.**

91°-179°

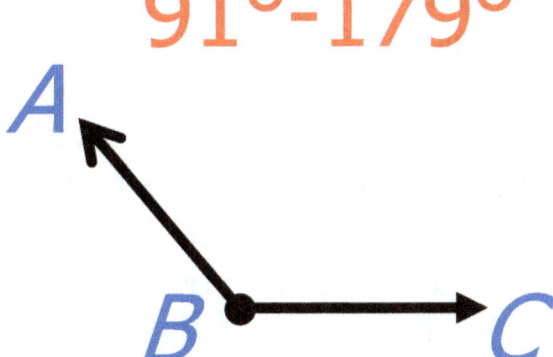

If this angle measures 130°, we would use the notation shown below to record its measurement.

$$m\angle B = 130°$$

The measure of angle B is 130°.

Drawing and Measuring Angles

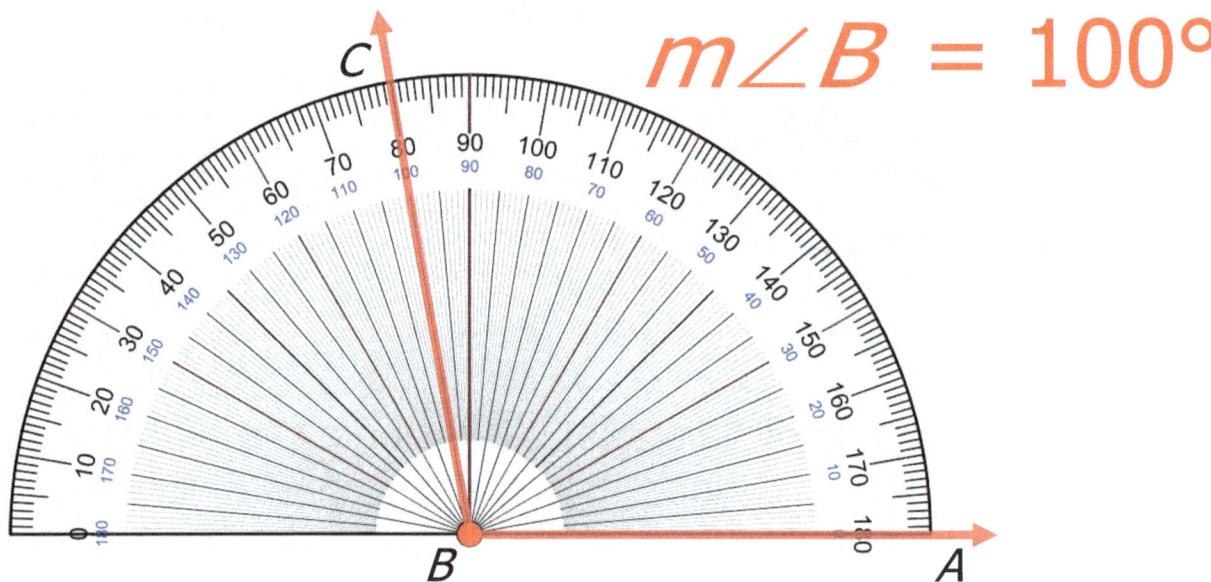

$m\angle B = 100°$

What is the measurement of the angle? Since the bottom line is positioned on the right side of the protractor, we will be looking at the inner (or bottom) number where the other arrow is pointing to determine its measurement. (The inner number of this angle is 100°.) Note that the outer (or top) measurement reads 80°, but you can tell plainly that this is not an 80° angle because it is wider than a right angle, which is 90°.

How to use a protractor to draw an angle: A protractor can be used to draw and measure various angles. To draw an angle facing the same position as that above, use the bottom of your protractor to draw a line on your paper. Position the protractor over the line you drew so that the center point of the protractor (marked *B* in this example) is at one end of your line and the other end lines up with the 0° mark on the right side (marked *A*). Count up from zero and put a point above the number of degrees you want for your angle (100° in this example). Again, since the bottom line is positioned on the right side of the protractor, we will be looking at the inner (or bottom) numbers of it. This point can be named *C*, and then the protractor would be used to draw a line from *B* to *C*.

Drawing and Measuring Angles

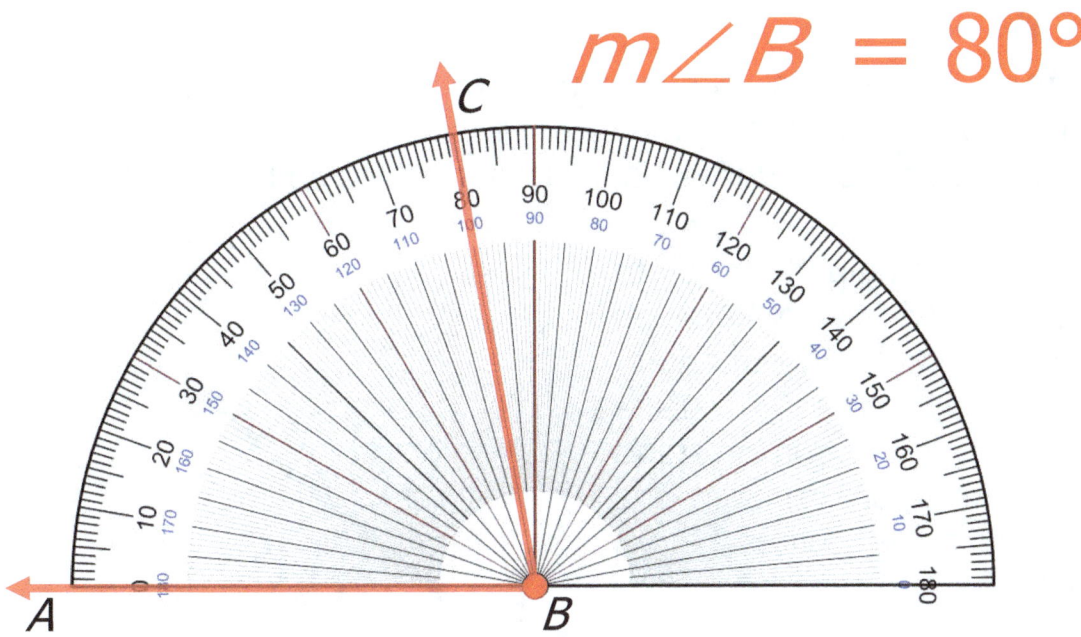

$m\angle B = 80°$

What is the measurement of the angle? Since the bottom line is positioned on the left side of the protractor, we will be looking at the outer (or top) number where the other arrow is pointing to determine its measurement. (The outer number of this angle is 80°.) Note that the inner (or bottom) measurement reads 100°, but you can tell plainly that this is not a 100° angle because it is not even as wide as a right angle, which is 90°.

Teacher instructions: Follow the instructions on the previous page to draw an angle and let students identify its measurement and the type of angle it is (i.e., right, acute, obtuse, straight, or a zero angle).

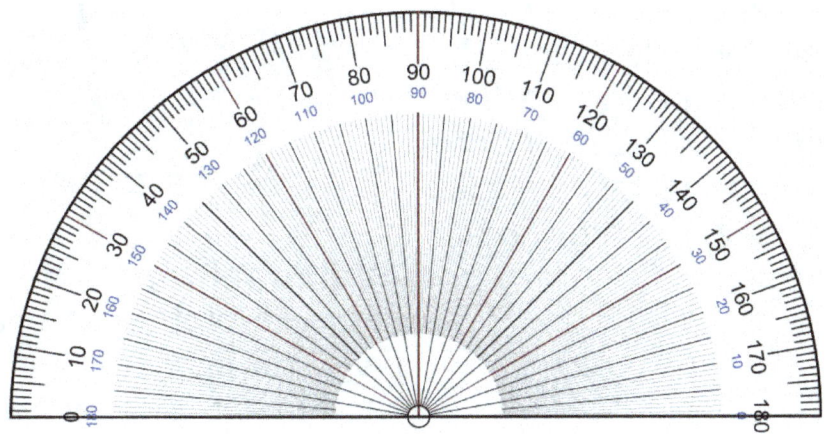

Complementary 90°

Instruction: Two angles that measure 90° when added together are complementary. The complement of a 50° angle is 40° because $50 + 40 = 90$. Recall that 90° is also what a right angle measures.

(**Hint:** It might be helpful for you to remember the difference between complementary and supplementary this way. *Complementary* starts with *c*, and *supplementary* starts with *s*; *c* comes before *s* in the alphabet, just as *90* comes before *180* in numbers.)

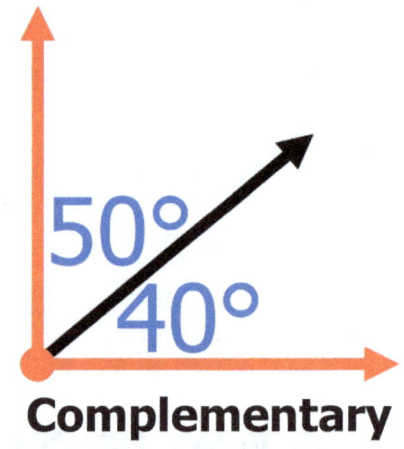
Complementary

Supplementary 180°

Instruction: Two angles that measure 180° when added together are supplementary. The supplement of a 100° angle is 80° because $100 + 80 = 180$. Recall that 180° is also what a straight angle measures.

The 100° angle and the 80° angle form a **linear pair** because they are adjacent (side by side), and the sides they don't share form a line. (The word *line* is in *linear*.) Notice that the complementary angles above are adjacent, but they do not form a linear pair.

Supplementary

Additional practice: You can use the lower protractor on the previous page to find the complement or supplement of various angles.

Three times the complement of angle A exceeds the supplement of angle A by 48°. What is the measure of angle A?

$$A = 21°$$

The complement of angle A: $90 - A$
The supplement of angle A: $180 - A$
$3(90 - A) = 180 - A + 48$
$270 - 3A = 228 - A$
$42 = 2A$, $A = \mathbf{21°}$

Check your answer: We know that 21 is the correct value of A because both of the original equations equal 207.
 $3(90 - 21) = \underline{207}$ and
 $180 - 21 + 48 = \underline{207}$.

If the complement of angle θ is 40° less than half the supplement of angle θ, what is the measure of angle θ?

$$\theta = 80°$$

$$90 - \theta = \tfrac{1}{2}(180 - \theta) - 40$$
$$90 - \theta = 90 - \tfrac{1}{2}\theta - 40$$
$$90 - \theta = 50 - \tfrac{1}{2}\theta$$
$$40 = \tfrac{1}{2}\theta,\ \theta = \mathbf{80°}$$

Three times the complement of angle θ equals 210°. What is the measure of angle θ?

Instruction: Recall that two angles that measure 90° when added together are complementary. Thus, the complement of angle θ could be recorded as $(90 - \theta)$. When we add the other information from the word problem, we have an equation that reads:

'$3(90 - \theta) = 210°$. Now we can solve for θ.

$$\theta = 20°$$

$$'3(90 - \theta) = 210°$$
$$3 \times 90 = 270, \; 3 \times -\theta = -3\theta$$
$$270 - 3\theta = 210$$
$$210 - 270 = -60$$
$$-3\theta = -60$$
$$-60 \div (-3) = 20$$

Check your answer: We know that *20* is the correct value of θ because if you replace θ in the original equation with 20, the answer is 210.

$$'3(90 - 20) = 210°$$

Exterior and Interior Angles

Instruction: Interior angles are inside a shape, while exterior angles are outside a shape. The 105° angle in the figure below is an exterior angle. Notice that the exterior angle forms a linear pair (a straight line) with an angle of the triangle. This means that an exterior angle will be supplementary to the angle it is adjacent to (together they will measure 180°--the same as a straight angle). For example, if the angle adjacent to the exterior angle measures 75°, then the exterior angle measures 105°. Find the measure of the other exterior angles.

Another way you can find the measurement of an exterior angle is to add the **remote interior angles** together. In the figure below, the remote interior angles of the exterior angle 105° are the 55° and 50° angles because they are not adjacent to it (55 + 50 = 105). You can illustrate the exterior angles of a triangle by extending each side beyond the vertex.

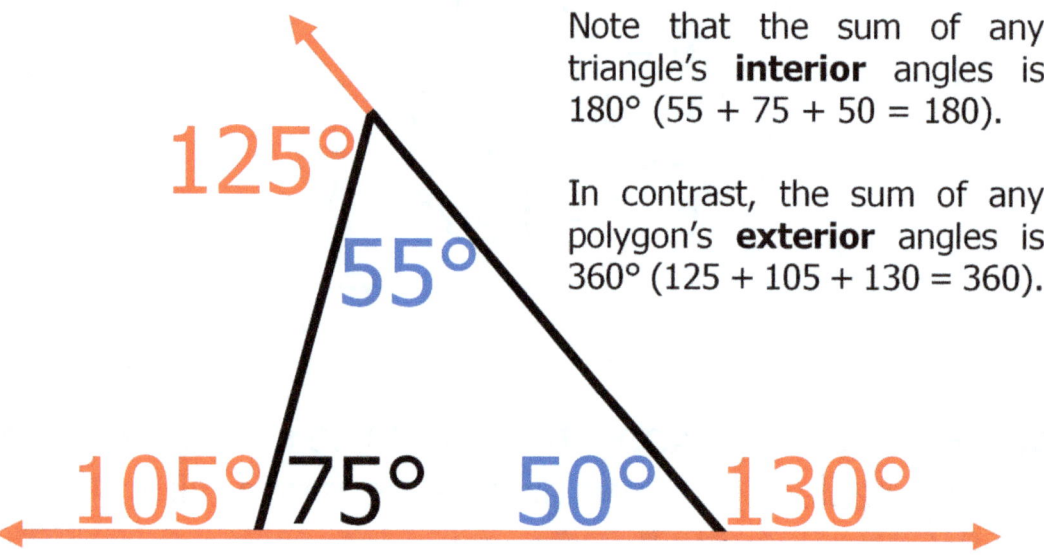

Note that the sum of any triangle's **interior** angles is 180° (55 + 75 + 50 = 180).

In contrast, the sum of any polygon's **exterior** angles is 360° (125 + 105 + 130 = 360).

Angle Bisectors

Instruction: The **perpendicular bisector** in the straight angle below is $\overrightarrow{BD}$. If you bisect a 180° angle (or cut it in half), then the two new angles measure 90° because 180 ÷ 2 = 90.

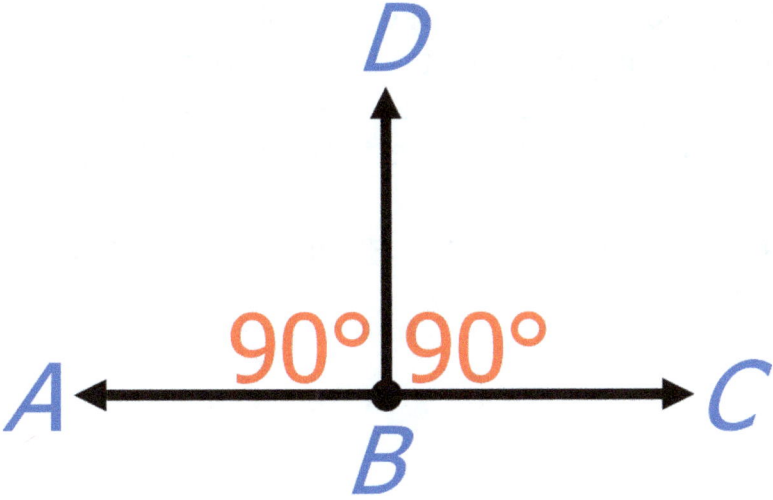

The **angle bisector** of this right angle is $\overrightarrow{BD}$. If you bisect a 90° angle (or cut it in half), the two new angles will measure 45° because 90 ÷ 2 = 45.

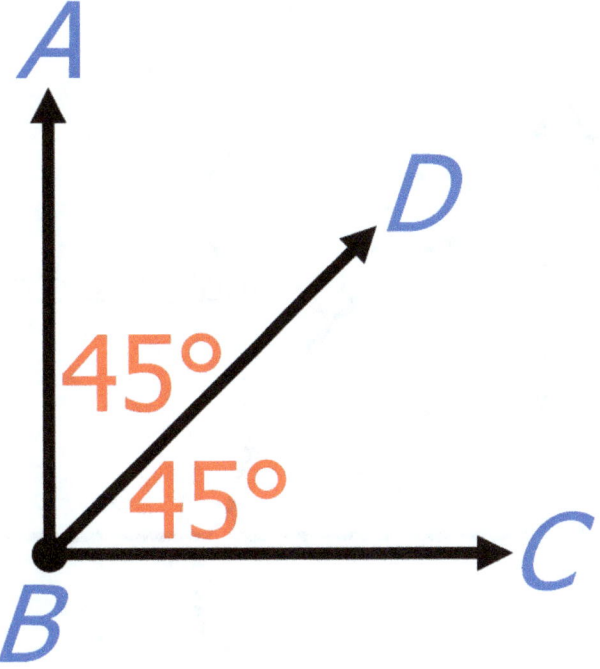

Angle Bisector

1. Use your protractor to draw a right angle and name it ∠EFG. (Put your protractor away.)
2. With the compass needle on point F, make an arc that intersects both sides of the angle (see the illustration). On each side, put a point where the arc intersects the angle's side and name the points H and I.
3. Put the compass tip on H and make a small arc that is centered between both sides of the angle (adjust your compass as needed). Without changing the setting you used, put the compass needle on point I and make a second small arc to intersect the first. Name the intersection of the two arcs J.
4. Use your ruler to draw an <u>angle bisector</u> from F through point J.

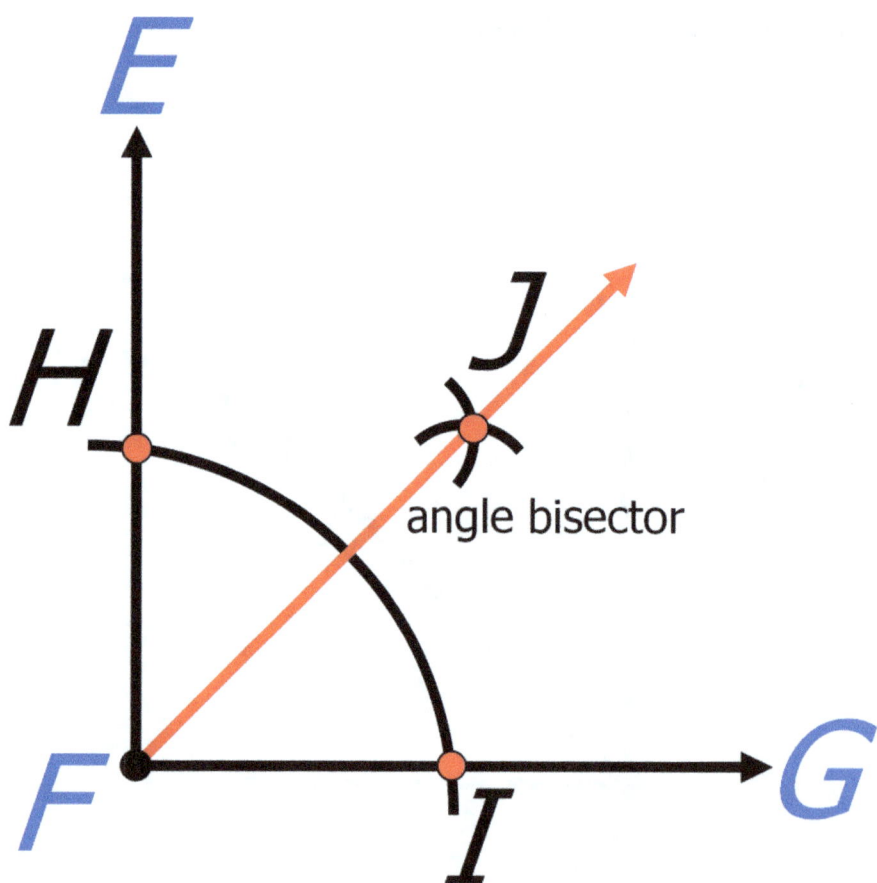

Perpendicular Bisector

1. Use your protractor to draw a line segment and name it $\overline{EG}$.
2. Open your compass so that it is a little more than half the length of the line segment. (Do not change this setting.) With the compass tip on point E, make an arc through the line segment. Then, put the compass on point G and make a second arc through the line segment.
3. Use your ruler to draw a <u>perpendicular bisector</u> from the point above the line segment where the two arcs intersect to the point below the line segment where the two arcs intersect.
4. Put a point to show where the perpendicular bisector intersects the line segment and name it F; F is the **midpoint** of $\overline{EG}$.

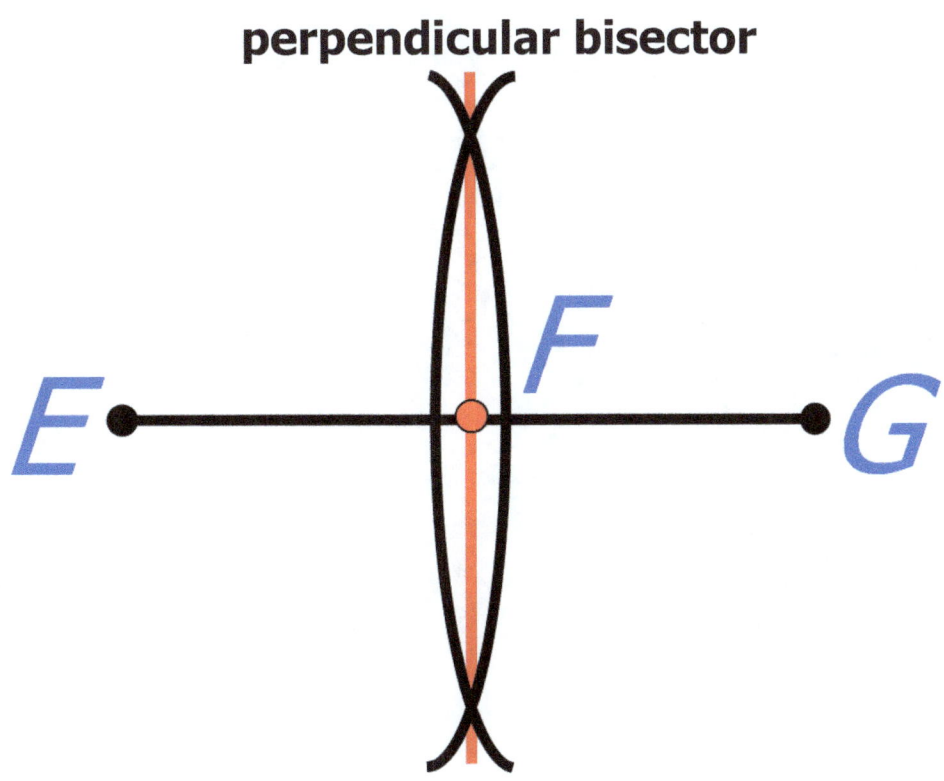

55

Construct a line segment that is congruent to $\overline{YZ}$.

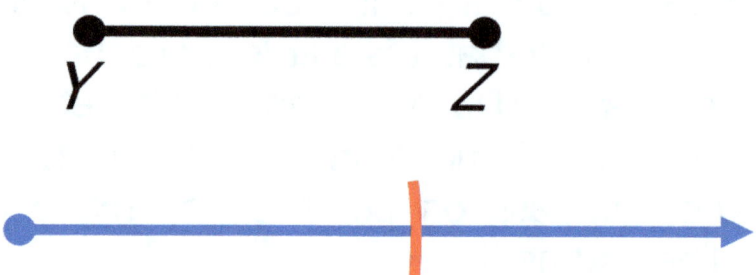

Instructions: Put the point of your compass on *y* and open it so that the pencil extends to point *z*; leave your compass in this position while you draw a ray (see the bottom figure). Put the compass needle on the end point of the ray and make an arc through the ray.

Draw a line through P that is perpendicular to $\overline{AB}$.

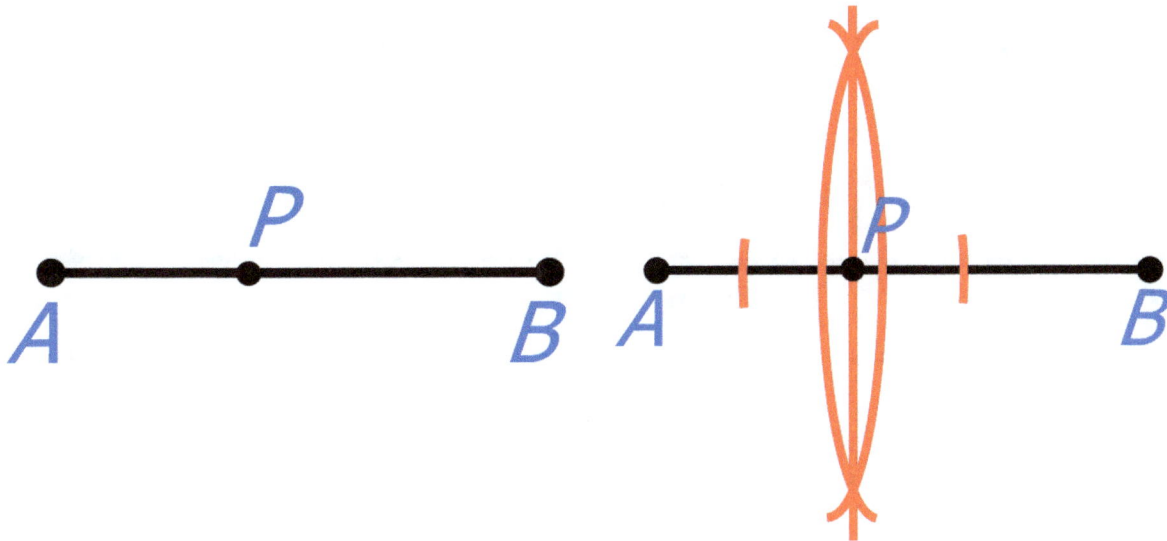

Instructions
1. Put the tip of the compass on P and adjust the setting to make an arc somewhere between A and P. Then, without changing the setting, make an arc between P and B.
2. Put the tip of your compass on the first arc. The compass should be adjusted to go slightly past P from this position. With the compass tip on the first arc, make a larger arc through the line segment. Then, without changing the setting, put the compass on the second smaller arc and make another larger arc through the line segment.
3. Use your ruler to draw a <u>perpendicular bisector</u> from the point above the line segment where the two arcs intersect to the point below the line segment where the two arcs intersect.

Draw a line through *P* that is perpendicular to $\overline{FG}$.

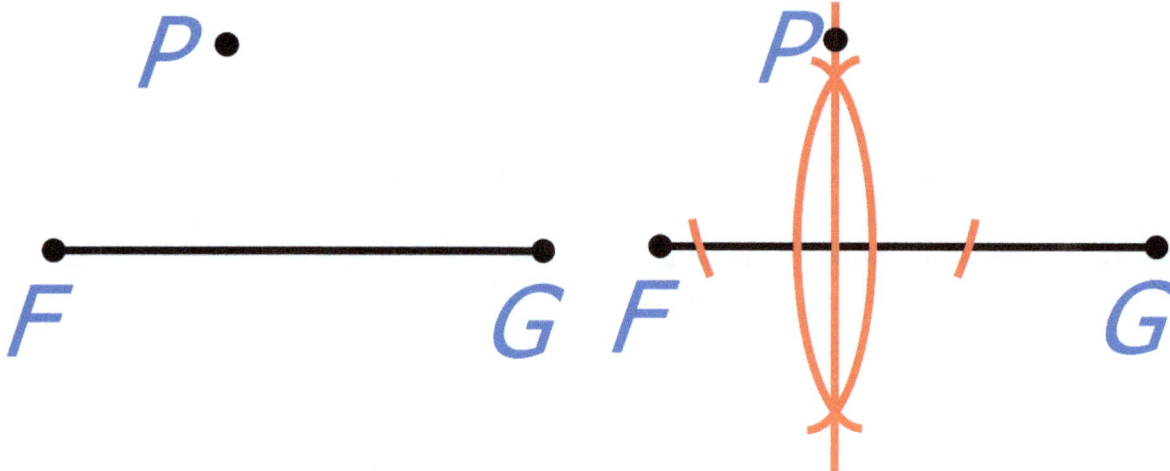

Instructions: Put the tip of the compass on point *P* and adjust the setting to make an arc somewhere on line *FG* between *F* and point *P*. Then, without changing the setting, make an arc on the line segment between point *P* and *G*. Finally, use the procedure from the previous page to construct the perpendicular bisector of the two arcs.

Construct a right triangle from the given line segments. The hypotenuse must be the same length as *n* and one of the legs the same length as *m*.

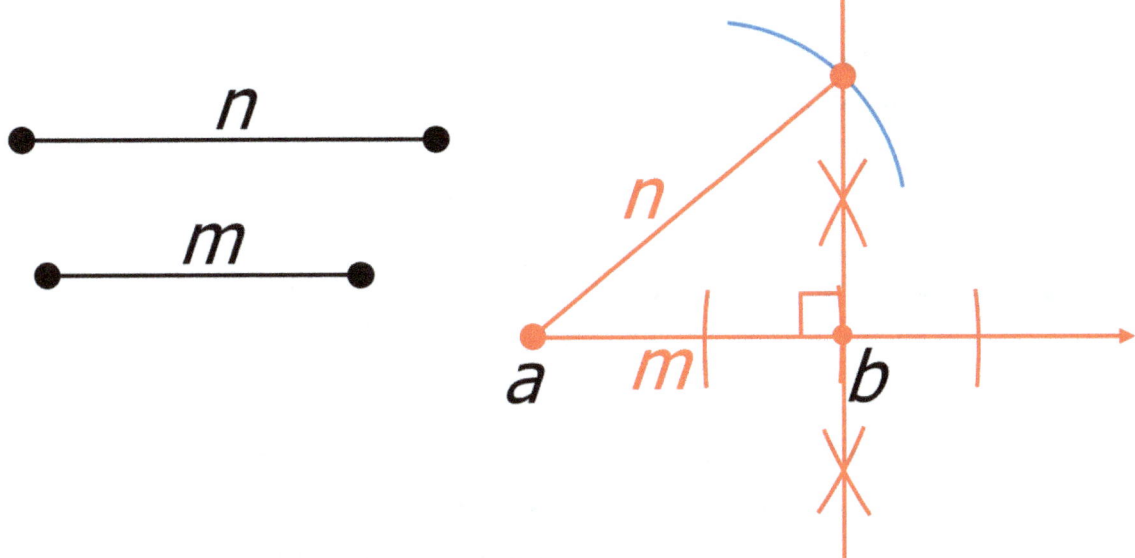

Instructions: To create a right triangle from the line segments given, begin by drawing a ray. The end point of the ray will be the first vertex of the triangle, which we will name *a*. Next, set your compass so that it is the length of the shorter line segment, which in this example is *m*. Put the point of your compass on *a* and draw an arc through the ray. (The point at which the arc intersects the ray will be the second vertex of the triangle, which we will name *b*.)

Now, set your compass so that it is the length of the longer line segment, which will be the hypotenuse of the right triangle. Put the point of your compass back on *a* and draw an arc above the ray (see the blue arc). Following the instructions from the page before last, draw a line through *b* that is perpendicular to $\overline{AB}$. Complete your right triangle by inserting the hypotenuse.

Create a triangle whose sides are congruent to x, y, and z.

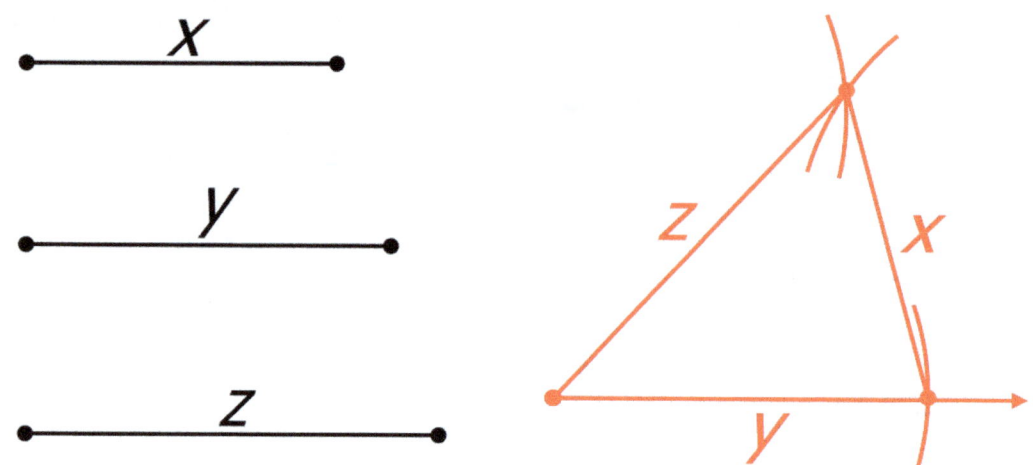

Instructions: To create the triangle from the line segments given, begin by drawing a ray. Next set your compass so that it is the length of x, y, or z. (This example uses y.) Put the compass needle on the end point of the ray (which will be the first vertex of the triangle) and draw an arc through the ray. (The point at which the arc intersects the ray will be the second vertex of the triangle.)

Set your compass so that it is the length of, let's say, z. Put the point of your compass back on the end point of the ray and draw an arc above the ray. Finally, set your compass so that it is the length of x. Put the tip of your compass on the point of the ray intersected by the arc and draw another arc above the ray that intersects the first one. The point where these arcs intersect is the third vertex of the triangle.

Construct an acute angle congruent to ∠ABC.

Instructions: Draw a ray and name it *MN* or any other letters of your choice. (Note that it doesn't have to be the exact same length as ray *BC* since this is not the line that will affect the measurement of the angle.)

Put the point of the compass on *B* and draw an arc through ∠ABC (see the blue arc in the first box). Consider the two points at which the arc intersects ∠ABC and name the points *Y* and *Z*. Now move the compass needle to *M* (being careful not to change the setting) and draw an equal arc through ray *MN*. The point at which the arc intersects the ray can be named *K*.

Move the point of the compass back to the original angle and put the point on *Y*. Adjust the compass so that it is the length of arc *YZ*. A small arc might be made at this point. Now move the compass back to the congruent angle you are constructing and put the point on *K*. Being careful not to change the compass setting, make an arc above the ray that intersects your first arc. Finally, use a ruler to draw ray *ML*.

Construct a line that goes through point P and that is parallel to $\overline{AB}$.

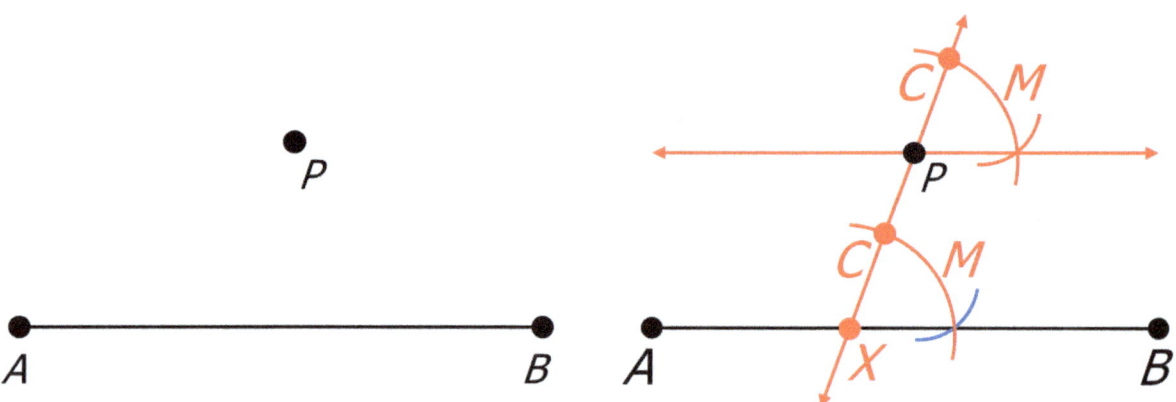

1. Draw a line that goes through point P and $\overline{AB}$. Name the point where the line intersects $\overline{AB}$. (In this example, the point is named X.) Note that the exact point where it intersects line segment AB is not important.
2. From point X, use your compass to draw an arc through $\overline{AB}$ and the line that goes through point P (see the bottom red arc). This arc represents the angle that has been formed by the intersecting lines. Name the angle (it is named M in this example). Move the compass needle to point P, being careful not to change the setting, and copy the arc. Name the points where each arc intersects the line through point P (we have named them C).
3. Put the tip of the compass on point C above X. From this position, make a small arc that intersects the arc representing the angle and $\overline{AB}$ (see the blue arc). Now, move the tip of the compass to point C above P, being careful not to change the setting, and copy the arc.
4. Finally, draw a line that goes through point P and the intersecting arcs.

Notes

Divide line segment AB into three congruent sections. (Use construction.)

Instructions: First, draw ray AC so that it forms an angle with $\overline{AB}$. Put the compass needle on point A and make an arc no more than 1/3 of the way through ray AC. Name the point P. Put the point of your compass on P (being careful not to change the setting) and make another arc through ray AC. Name the point Q. Next, put the point of your compass on Q and make a third arc through ray AC. Name the point R. Use your ruler to draw a line from point B to R.

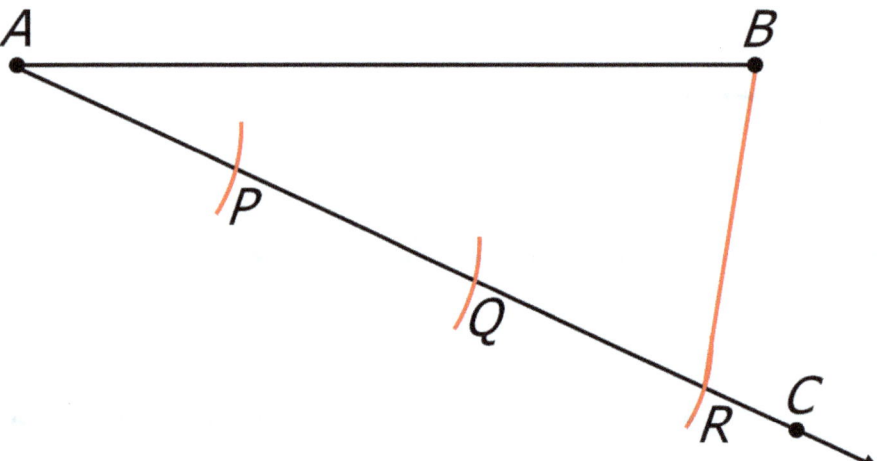

Put the point of the compass on R, adjust the compass to an appropriate setting, and make a large arc through ray AC and $\overline{BR}$ as shown below. Next, move the tip of the compass to Q (being careful not to change the setting) and make an equivalent arc at point Q. Repeat at point P.

Set the tip of your compass on the point of ray AC marked in this illustration with a blue circle. From this point, make a smaller arc that goes through the point where the larger arc and line segment BR intersect (see the green arc). Repeat at the other larger arcs, being careful not to change the compass's setting.

Finally, use your ruler to draw a line from point P to line segment AB. The line must go through the intersecting arcs. Repeat at point Q.

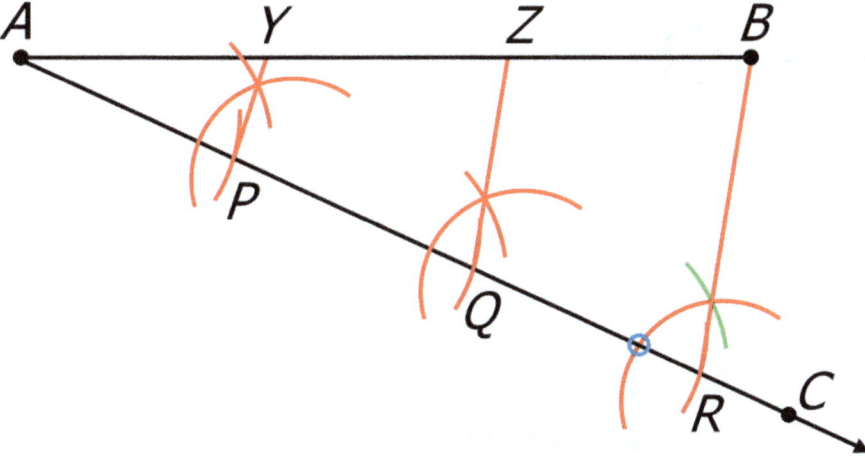

(Note that this same procedure could be used to divide $\overline{AB}$ into two equal sections, into four equal sections, or into other congruent sections specified by the directions.)

Finding the Measure of Bisected Angles

Instruction: Recall that a straight angle measures 180° and that if you bisect a 180° angle (or cut it in half), then the two new angles measure 90° because 180 ÷ 2 = 90. Notice from the figure at the bottom of the page that if you cut one of the 90° angles in half, you form two 45° angles, but the three angles would still equal 180° when added together (90 + 45 + 45 = 180).

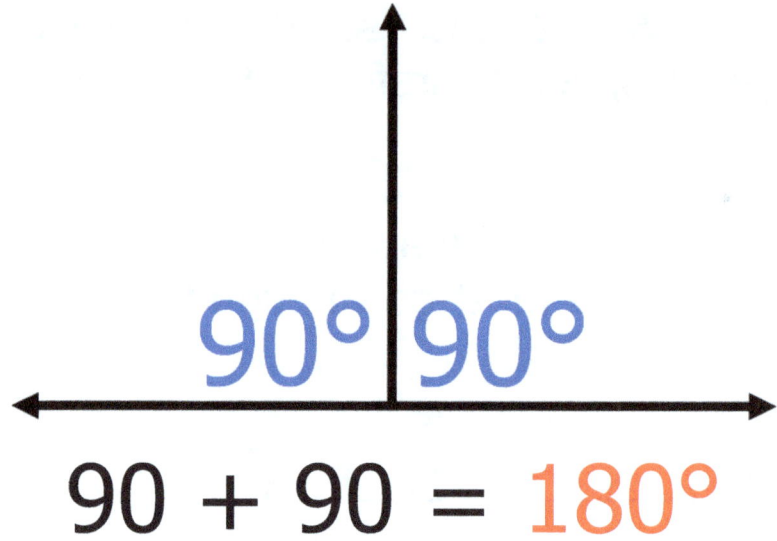

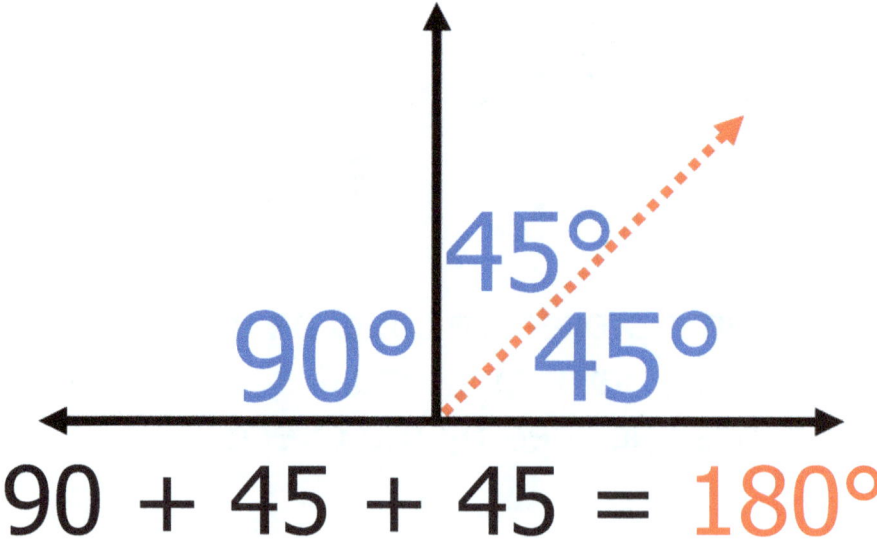

In the figure below, *a* is a straight line. If angle *x* measures 72°, what is the degree measure of *y*?

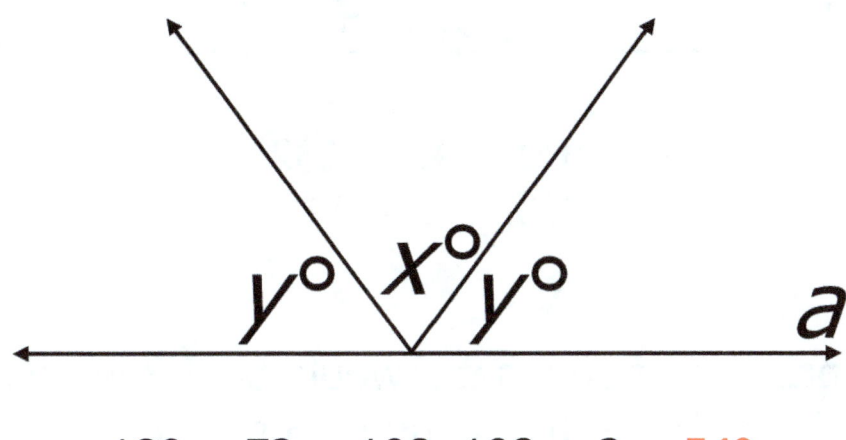

180 − 72 = 108, 108 ÷ 2 = 54°

Instruction: Since the two angles we are looking for are both symbolized by the variable *y*, they have to be the same measure and different from the *x*° angle.

Find the value of *x*.

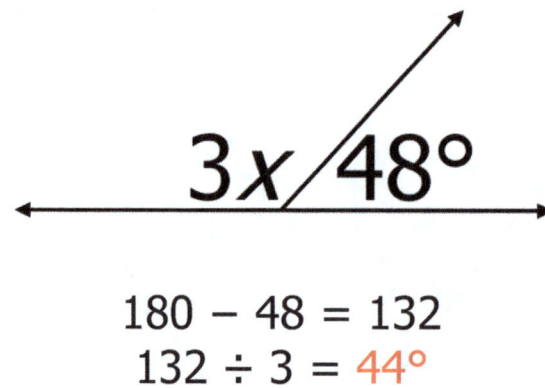

$180 - 48 = 132$
$132 \div 3 = 44°$

Instruction: First, find what *x* would be if 3 were not beside it. Then divide that value by 3 to find *x*.

Find 2*x*.

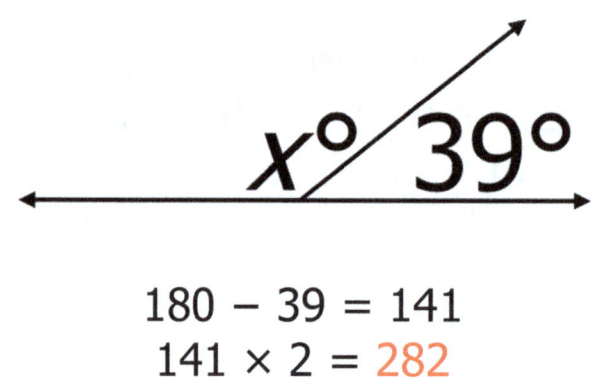

$180 - 39 = 141$
$141 \times 2 = 282$

Instruction: Since we are looking for 2*x*, find the measure of *x* and then multiply it by 2.

alternate interior angles and alternate exterior angles

Instruction: In the figure below, *t* is a **transversal**. A line that intersects two or more parallel lines in the same plane is a transversal. When two parallel lines are cut by a transversal, the *alternate interior angles* are congruent (they have the same measure). Alternate exterior angles also have the same measure. Alternate interior angles are inside the parallel lines, and alternate exterior angles are outside the parallel lines. In the figure, 2 and 7 are alternate interior angles, and 6 and 3 are alternate interior angles; 1 and 8 (or 5 and 4) are alternate exterior angles.

Proving two lines are parallel: The two black lines that are intersected by the transversal are parallel. You can prove that they are parallel by showing that the <u>alternate interior angles</u> (or the <u>alternate exterior angles</u>) are congruent. If they are not congruent, then the lines are not parallel.

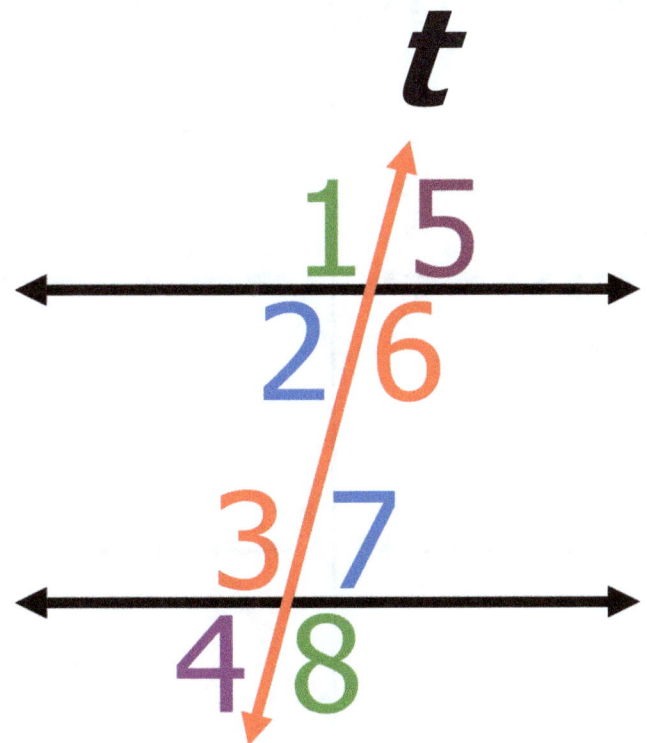

Corresponding and Vertical

Instruction: Corresponding angles have the same measure. In the first figure below, corresponding angles include 1 and 3, 2 and 4, 5 and 7, and 6 and 8. If you cut the transversal in half and put one of the parallel lines on top of the other, corresponding angles would be in the same position.

Vertical angles form an *x*. In the second figure below, angles 1 and 6, 5 and 2, 3 and 8, and 7 and 4 are vertical angles. Vertical angles have the same measure (if angle 1 measures 105°, then angle 6 also measures 105°).

Corresponding | Vertical

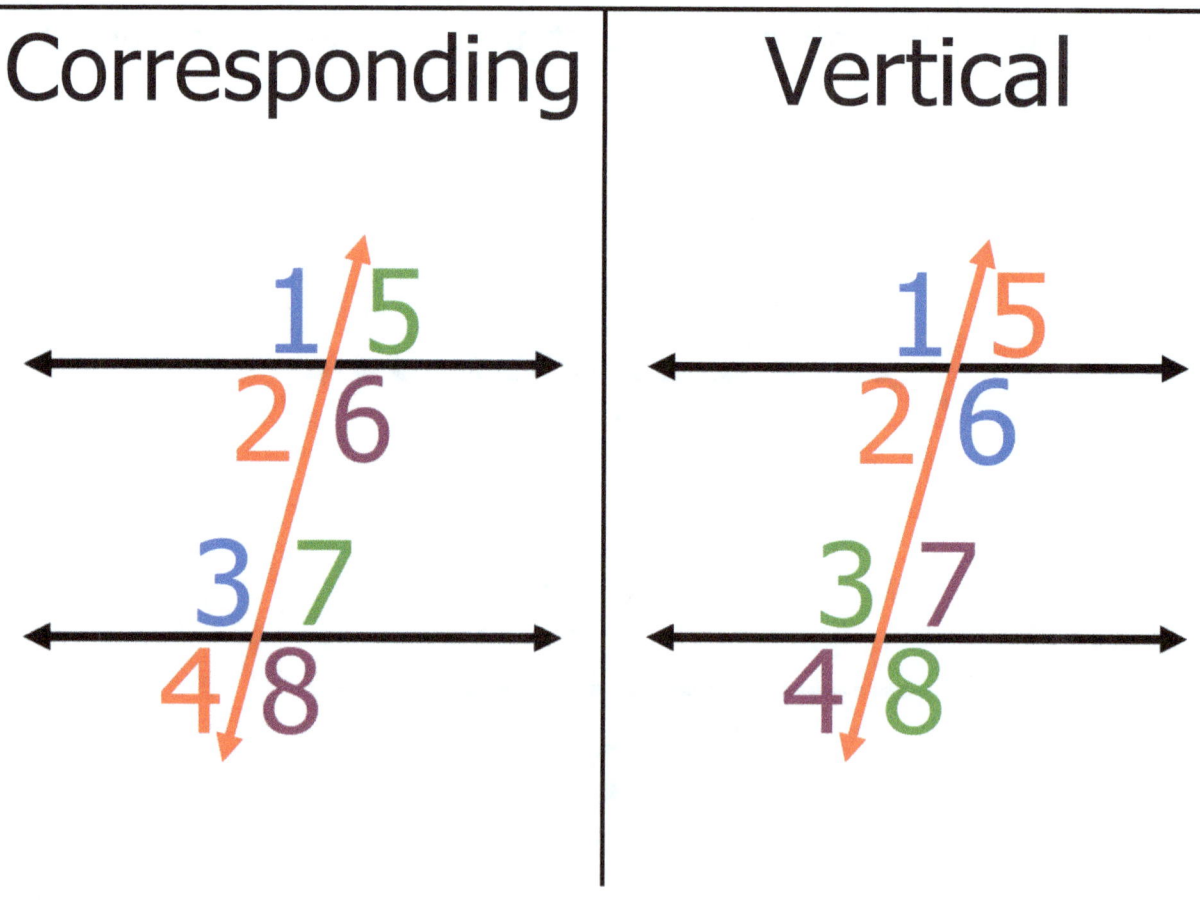

Adjacent Angles

Instruction: <u>Adjacent angles</u> are side by side; they are formed by two intersecting lines. In the figure below, 1 and 5 are adjacent angles. Other adjacent angles include 2 and 6, 3 and 7, and 4 and 8. Additionally, 1 and 2, 5 and 6, 3 and 4, and 7 and 8 are adjacent angles. Together, adjacent angles measure 180°—the measure of a straight angle. Therefore, if angle 1 measures 105°, then angle 5 measures 75° because 105 + 75 = 180.

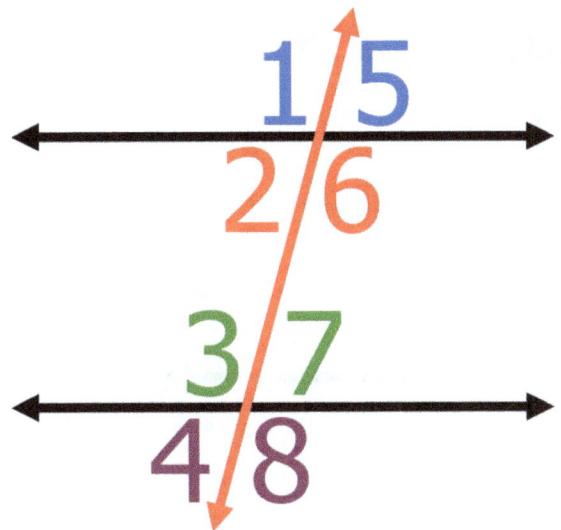

The two lines intersected by the transversal in the figure to the left are parallel. Thus, the two interior (or inside) angles on the same side of the transversal will equal 180° when added together (e.g., angles 2 and 3 will equal 180°, and angles 6 and 7 will equal 180°).

Comparably, the two exterior (outside) angles on the same side of the transversal equal 180° when added together (e.g., angles 1 and 4 will equal 180°, and angles 5 and 8 will equal 180°).

Instruction: In review, <u>alternate interior angles</u> have the same measure, <u>alternate exterior angles</u> have the same measure, <u>corresponding angles</u> have the same measure, <u>vertical angles</u> have the same measure, and <u>adjacent angles</u> are supplementary (together they measure 180°).

Practice: Choose an appropriate measure for one of the angles and use it to identify the other angle measurements.

Practice: Number the angles and then identify any alternate interior angles, alternate exterior angles, corresponding angles, vertical angles, or adjacent angles.

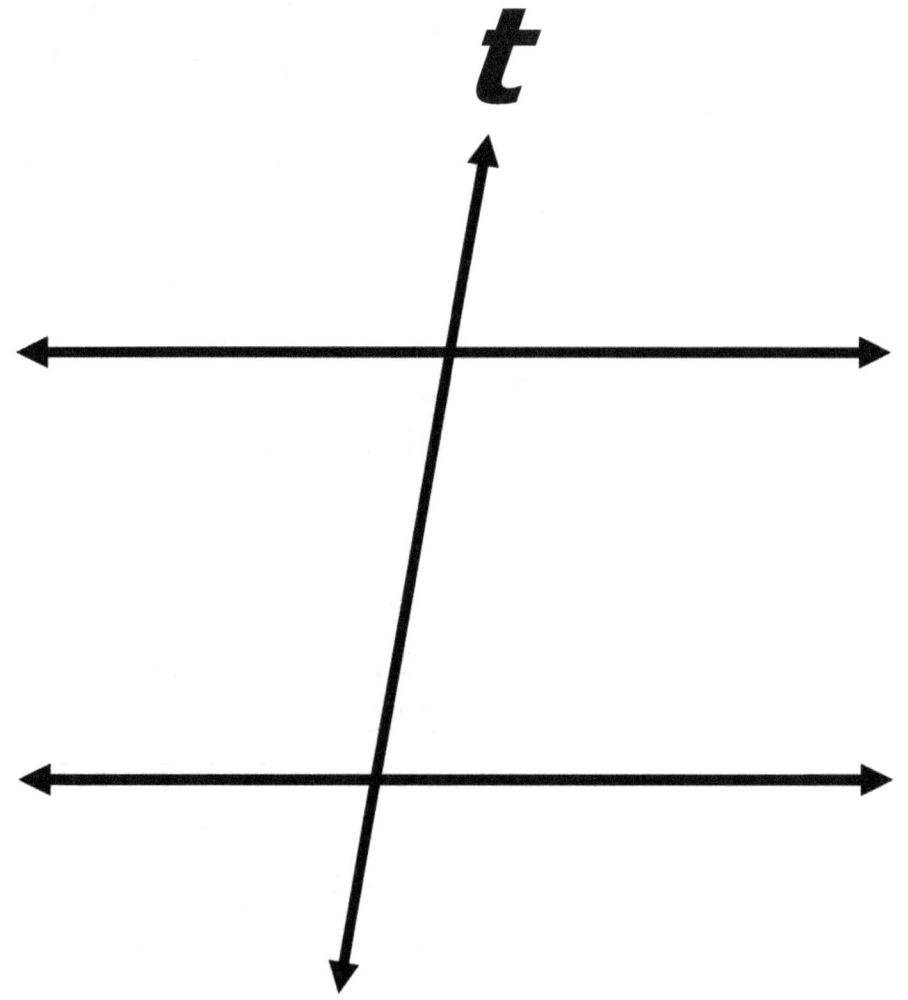

Find a, b, c, and d.

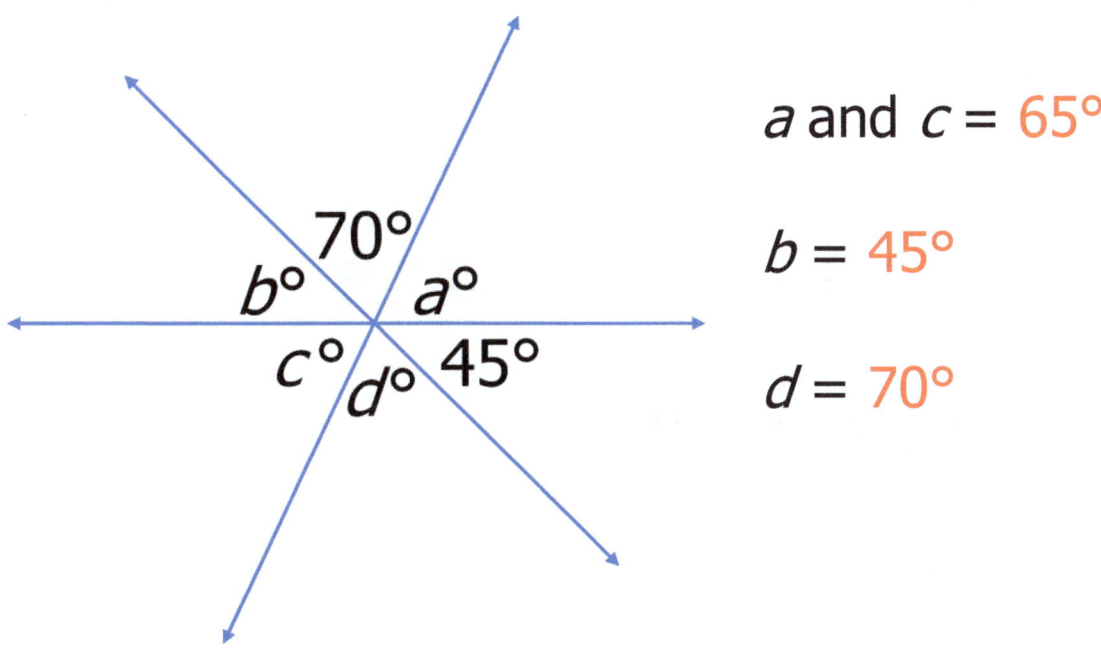

a and c = 65°

b = 45°

d = 70°

b + 70° + a = 180° (because it forms a straight line)
c + d + 45° = 180° (because it forms a straight line)

Since the 45° angle and *b* are vertical angles, **b** also measures 45°.
Since the 70° angle and *d* are vertical angles, *d* also measures 70°.
Now that we know b = 45° and d = 70°, we can rewrite the equations.
   45° + 70° + a = 180°
   c + 70° + 45° = 180°

Solve for *a*: 180 − 45 − 70 = 65

The measure of *a* is 65°, and, since *a* and *c* are vertical angles, *c* also measures 65°.
   65° + 70° + 45° = 180°

If $y = 2x + 36$ and line *a* is parallel to line *b* (*a* ∥ *b*), what is the measure of angle *x* and of ∠*y*?

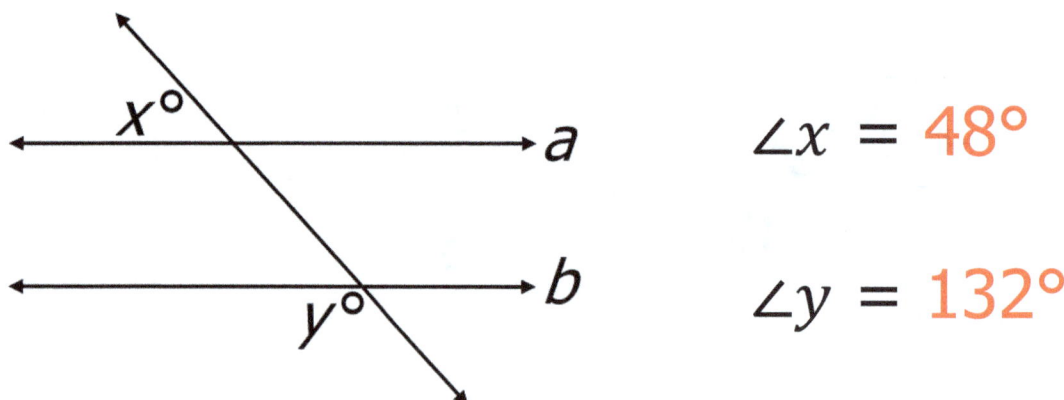

∠*x* = 48°

∠*y* = 132°

Instruction: Since line *a* is parallel to line *b* and angles *x* and *y* are two outside angles on the same side of the transversal, they have to equal 180° when added together.

    $x + y = 180°$

Replace *y* with $2x + 36$ since this is what the word problem tells us that *y* equals.

    $x + 2x + 36 = 180°$

Simplify by adding the *x* to 2*x* to get 3*x*; also subtract 36 from 180 to get 144.

    $3x = 144$

Divide 144 by 3 to find the value of *x* (144 ÷ 3 = 48).

    $x = 48°$

Replace *x* with 48 in $y = 2x + 36$, and you find that the value of *y* is 132.

    $2(48) + 36 = 132°$

In the figure below, the arrowheads show us that the opposite lines are parallel. Line *a* is parallel to line *b* (*a* ∥ *b*), and line *y* is parallel to line *z* (*y* ∥ *z*). The measure of one angle is given. Find the measure of the other 15 angles.

We know that the angle adjacent to the 120° angle is 60° because 120 + 60 = 180. From there, we can use what we learned about corresponding, vertical, and adjacent angles to find the measure of the other angles.

Notice that every angle in this illustration is either 60° or 120°. Also notice that the angles that look congruent are congruent.

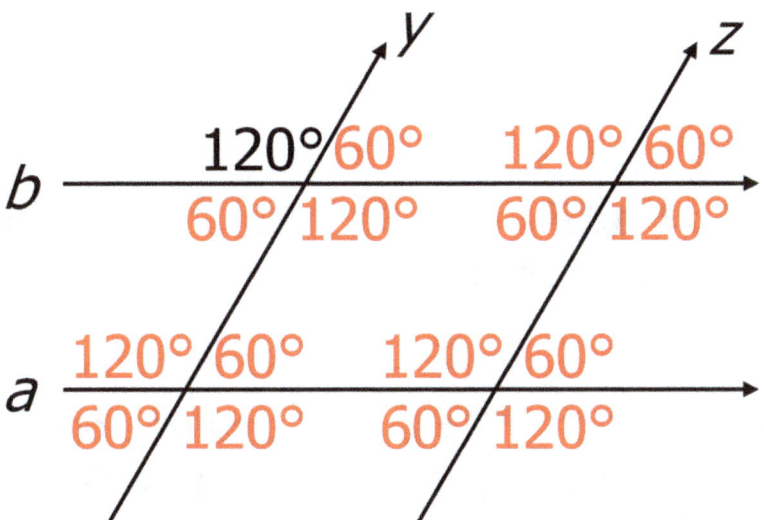

Find *n*.

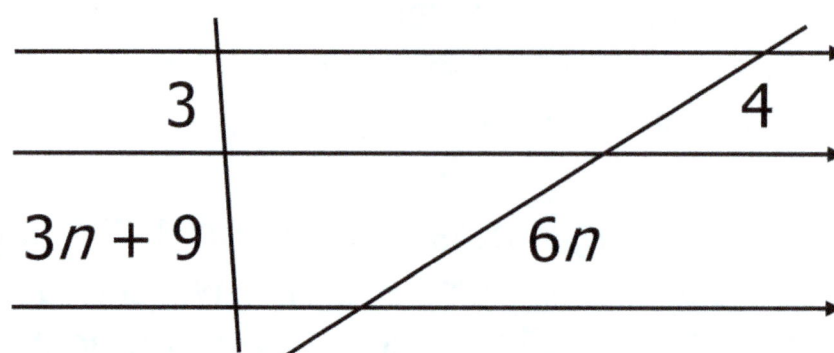

Solution:

$\dfrac{4}{6n} = \dfrac{3}{3n+9}$ $18n = 12n + 36$ $6n = 36$ $n = 6$

Instruction

Begin by writing the equation.

$$\dfrac{4}{6n} = \dfrac{3}{3n+9}$$

Then cross multiply. $3 \times 6n = \mathbf{18n}$ $4(3n + 9) = \mathbf{12n + 36}$
After cross multiplying, the equation reads:
   $18n = 12n + 36$

Simplify the equation above by moving $12n$ to the same side of the equal sign as $18n$. (Since $12n$ is positive, the only way you can do this is by subtracting it from $18n$.) $18n - 12n = 6n$
Now the equation reads:
   $6n = 36$.

Our goal is to get *n* by itself on one side of the equal sign so that we know it's value. We have to move 6 to the same side of the equal sign as 36. Since $6n$ represents multiplication, we do this by dividing it by 36 ($36 \div 6 = 6$).
   $n = 6$

Parallelograms

Instruction: Opposite angles (and sides) of a parallelogram are parallel and congruent. *Congruent* means that they are exactly the same. Also, if you add the measure of consecutive angles (that is, if you add the measure of one interior angle of a parallelogram to the angle before or after it), the sum will be 180°. If you add together what all four of the interior angles measure, the sum will always be 360°. Since this definition also describes squares and rectangles, squares and rectangles are considered special types of parallelograms.

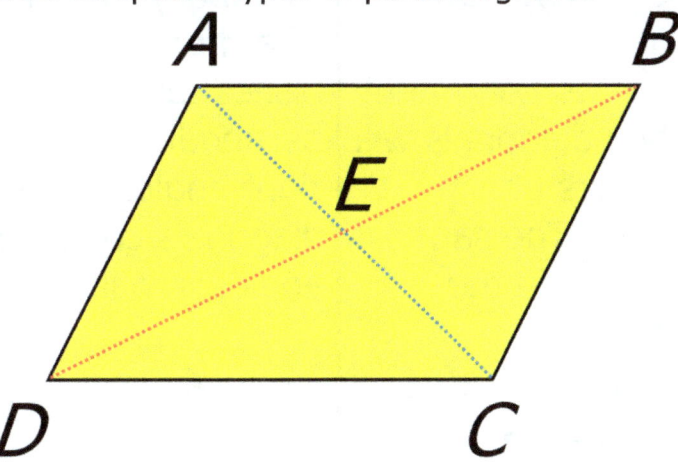

If you were asked to name a shape *ABCD*, you would record the letters in consecutive order around the shape. Although *A* could have been recorded on one of the other three corners of the polygon, it would still need to be followed by *B, C,* and *D*.

Put your finger on letter *A* and go around the shape counterclockwise to discover another name that could be used for the same illustration (*ADCB*). Put your finger on letter *B* and go clockwise around the shape to discover yet another name that could be used for this illustration (*BCDA*). Can you identify other possible names for this illustration?

In this parallelogram, examples of **consecutive angles** are ∠*A* and ∠*B* (or ∠*A* and ∠*D*). Consecutive angles are adjacent. (Can you find other consecutive angles?) **Opposite angles** include ∠*A* and ∠*C*. (Can you identify the other pair of opposite angles?) **Consecutive sides** include $\overline{AB}$ and $\overline{BC}$. (Can you identify other pairs of consecutive sides?) **Opposite sides** include $\overline{AB}$ and $\overline{CD}$. (Can you identify the other opposite sides?)

| **Parallelogram** | **Rhombus** |

Instruction: A diagonal of a flat shape connects the opposite vertices. The *diagonals* of a parallelogram are not the same length (unless it is a square or a rectangle, which are special types of parallelograms). One diagonal will divide a parallelogram into two congruent triangles, while two diagonals will divide it into two pairs of congruent triangles. The same is true of a rhombus. In the rhombus to the right, $\triangle ABE \cong \triangle CDE$ and $\triangle ADE \cong \triangle CBE$ ($\cong$ means "congruent"). Also, the red diagonal in the illustration bisects the blue diagonal and the blue diagonal bisects the red diagonal, so that $\overline{BE} = \overline{ED}$ and $\overline{AE} = \overline{EC}$.

Instruction: A rhombus is a parallelogram with all four sides the same length, although the diagonals of a rhombus are not the same length (unless it is a square, which is a special type of rhombus). When two diagonals of a rhombus (or of a square) bisect each other, they form perpendicular lines (lines with right angles). Also, the two diagonals cut the four vertex angles in half.

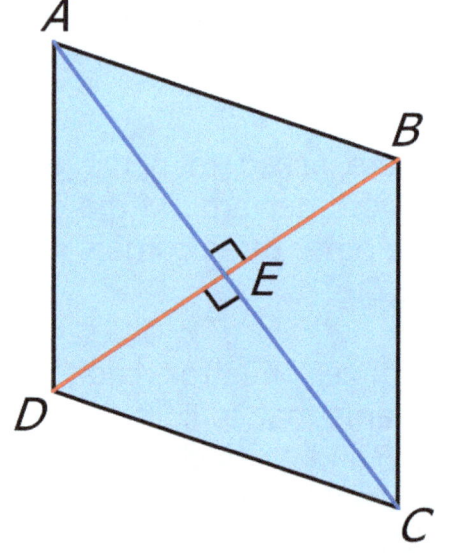

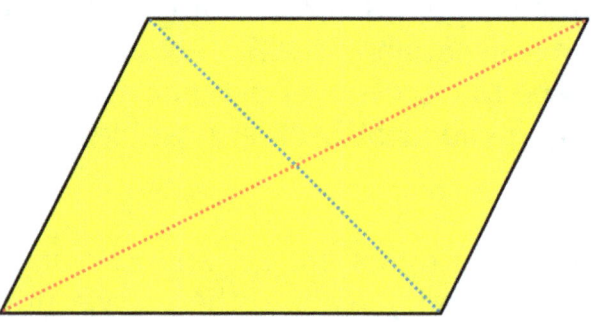

78

Rectangle	**Square**

Instruction: Opposite sides of a rectangle are parallel and congruent (*congruent* means that they are exactly the same), and all four of its angles are right angles. While one diagonal will divide a rectangle into two congruent right triangles, two diagonals will divide it into two pairs of congruent triangles. The diagonals of a rectangle are the same length. (In a parallelogram or rhombus, the diagonals are not the same length.) The red diagonal in the illustration bisects (cuts in half) the blue diagonal, and the blue diagonal bisects the red diagonal.

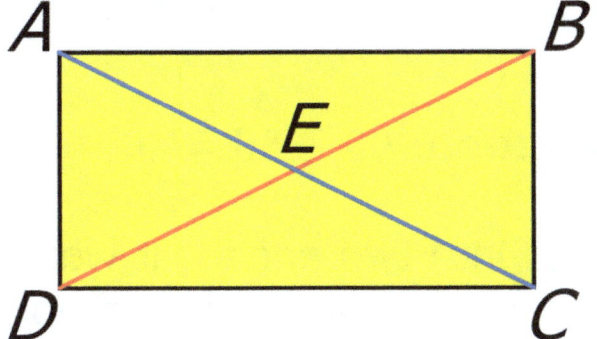

Instruction: A square has four right angles. One diagonal will divide a square into two 45°-45° right triangles, while two diagonals will divide it into four 45°-45° right triangles. Thus, two diagonals cut the four vertex angles of a square in half, creating eight 45° angles rather than four 90° angles. The diagonals are the same length. When two diagonals of a square (or rhombus) bisect each other, they form perpendicular lines (lines with right angles).

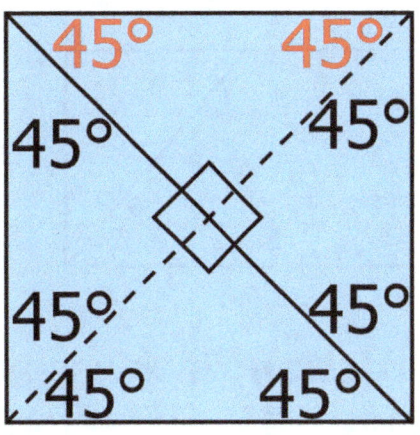

Find the measure of ∠CBD in the parallelogram to the right.

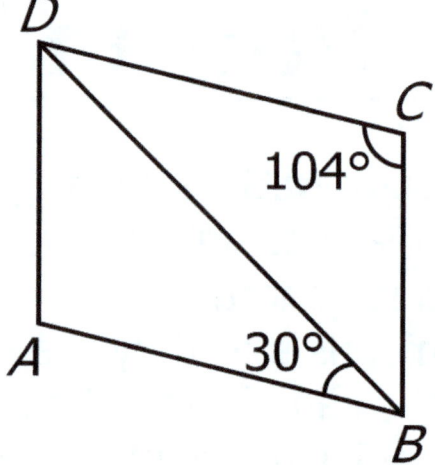

Solution: 180 − 104 = 76,
76 − 30 = 46°

Instruction: Recall that if you add the measure of one interior angle of a parallelogram to the angle before or after it, the sum will be 180°. Thus, we know that angles C and B will equal 180 without the diagonal. If you subtract the measure of angle c from 180, you have 76 (180 − 104 = 76). Since a diagonal is dividing angle b into two angles, subtract the measure of angle ABD from 76 (76 − 30 = 46°). Thus, the measure of angle CBD is 46°.

What is the measure of x in the rectangle below?

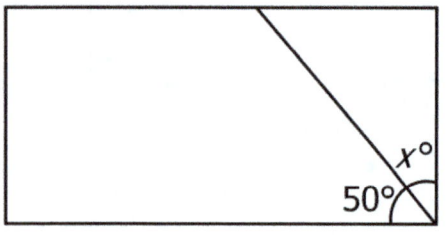   40° (**Solution:** 90 − 50 = 40)

Instruction: If x was not divided into two angles, it would measure 90°. Therefore, to find the measure of x, simply subtract the 50° angle from 90.

80

How many diagonals could you draw in a convex polygon with 6 sides?

9

The formula below is used to find the number of diagonals of a convex polygon. *N* stands for the number of sides the polygon has.

$$\frac{N(N-3)}{2}$$

$$\frac{6(6-3)}{2} = 9$$

$6 - 3 = 3, 6 \times 3 = 18, 18 \div 2 = \mathbf{9}$

Length of a Diagonal (Rectangular Solids)

$$d = \sqrt{(l^2 + w^2 + h^2)}$$

Instruction: To find the length of a diagonal in a rectangular solid, square the length, square the width, and square the height. Add the numbers you squared together. Then find the square root of the sum.

Problem 1: A rectangular solid has a length of 4 m, a width of 5 m, and a height of 2 m. What is the length of its diagonal?

$4^2 + 5^2 + 2^2 =$ ___, $16 + 25 + 4 = 45$, $\sqrt{45}$ m

Problem 2: Find the diagonal length of the rectangular prism. Use the radical sign in your answer.

$81 + 9 = 90$, $\sqrt{90}$, $\sqrt{90 + x^2}$ m

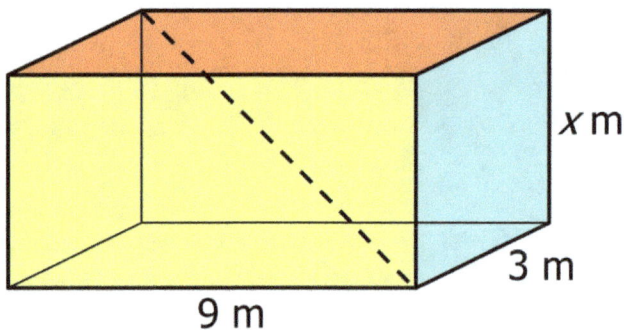

Length of a diagonal (Cube)

$$d = s\sqrt{3}$$

4 m

$4\sqrt{3} = 6.9$ m

Instruction: To find the length of a diagonal of a cube, multiply the length of one of its sides by the square root of 3. (Note that all the sides of a cube are equal.) If the length of its side is 4 m, plug *4*, *shift*, $\sqrt{}$, *3 EXE* (or *=*) into your calculator. In this example, $4\sqrt{3} = 6.92820323$, which is 6.9 when rounded to the nearest tenth.

The sum of a convex polygon's interior angles

Instruction: The formula "$(n - 2) \times 180$" is used to find what the total would be if you added all the interior angle measures of a convex polygon together (regular or not). You subtract 2 from the total number of angles the polygon has and then multiply the difference by 180.

With this formula, you can also find what each individual interior angle of a <u>regular polygon</u> measures or find an unknown interior angle measurement of a <u>polygon that is not regular.</u> In a regular polygon, all sides are the same length, and all angles are the same measure. (Examples of regular polygons include squares, equilateral triangles, or other shapes specified as regular, such as regular pentagons.) A square has four angles. Since $4 - 2 = 2$ and $2 \times 180 = 360$, you know that 360° would be the total if you added all the interior angle measures together.

$$(n - 2) \times 180$$

$$4 - 2 = 2,\ 2(180) = 360$$

If you wanted to know what each individual interior angle of this regular polygon measures, divide the total angle measure by 4 (the number of angles). Each angle of a square measures 90° because $360 \div 4 = 90$.

84

Instruction: If you need to know the sum from adding all the interior angle measures of a <u>convex</u> polygon (not concave) together, you can use the formula "**(*n* − 2) × 180**" to find that the sum will always be...

180° for a triangle   (3 − 2) × 180 = 180

360° for a quadrilateral   (4 − 2) × 180 = 360

540° for a pentagon   (5 − 2) × 180 = 540

720° for a hexagon   (6 − 2) × 180 = 720

900° for a heptagon   (7 − 2) × 180 = 900

1,080° for an octagon   (8 − 2) × 180 = 1,080

1,260° for a nonagon   (9 − 2) × 180 = 1,260

1,440° for a decagon   (10 − 2) × 180 = 1,440

The pentagon on the left is clearly not regular. If one angle measures 103°, a second measures 107°, a third measures 111°, and a fourth measures 119° (fill in the given measurements), what does the fifth angle measure?

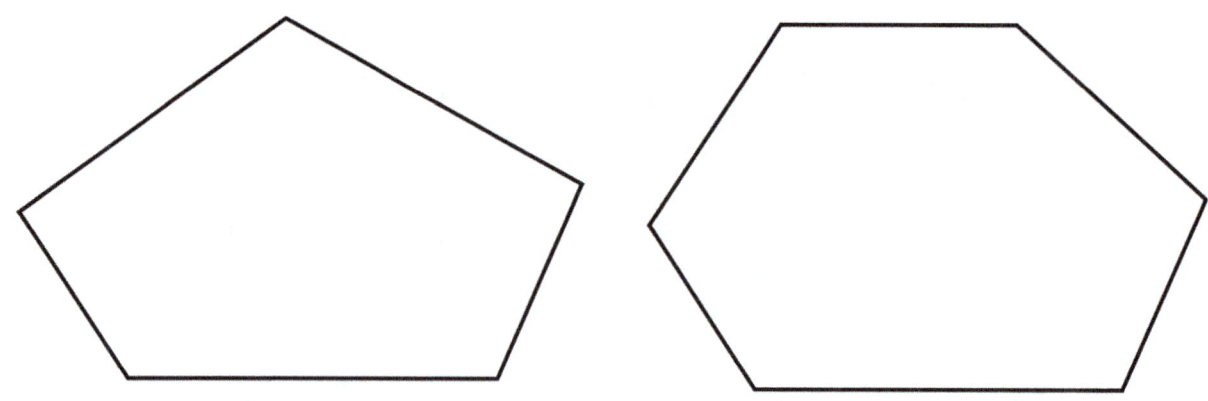

Instruction: To find the solution, first use the formula "$(n - 2) \times 180$" to determine what the total would be if you added all the angle measures of the polygon together. The total for the pentagon would be 540 because a pentagon has five sides and $(5 - 2) \times 180 = 540$. We can also calculate that the given angle measurements equal 440° when added together ($103 + 107 + 111 + 119 = 440$). When you subtract 440 from the total 540, you get 100 ($540 - 440 = 100$). Thus, the unknown angle measurement is 100°.

Teacher Instructions: The given measurements can be changed for additional practice, and you can make up your own problem for the hexagon in the second box to solve on the classroom screen.

Chapter 3

Angles and Triangles

(Suggested Grades: 8th and 10th)

Teacher instructions: Using *70 Times 7 Math: Electronic Textbook for Teachers (Geometry for Middle and High School Students),* ask students to identify any missing answers for you to write on the screen. Please note that since the answers are provided in student textbooks, they should have them closed during this time. Student textbooks can also be used as a key for the teacher's benefit.

Notes

Central Angles of Regular Polygons

Instruction: At the end of the previous chapter, we learned how to find the measure of each *interior angle* of a polygon. (In the square below, all four interior angles are 90°.) There is also a formula we can use to find the measure of each *central angle* of a regular polygon. The number of central angles of a regular polygon can be seen when you draw a line from the center of the polygon to each of its vertices. (The center of the polygon, represented by a black dot in the figure below, is the **vertex** of each central angle.) Central angles of a regular polygon are congruent. Use the formula below to find what each central angle of a regular polygon measures (divide 360 by the number of central angles, or sides, the regular polygon has).

$$360 \div n = \text{central angle}$$

$$360 \div \underline{}4\underline{} = \underline{}90°\underline{}$$

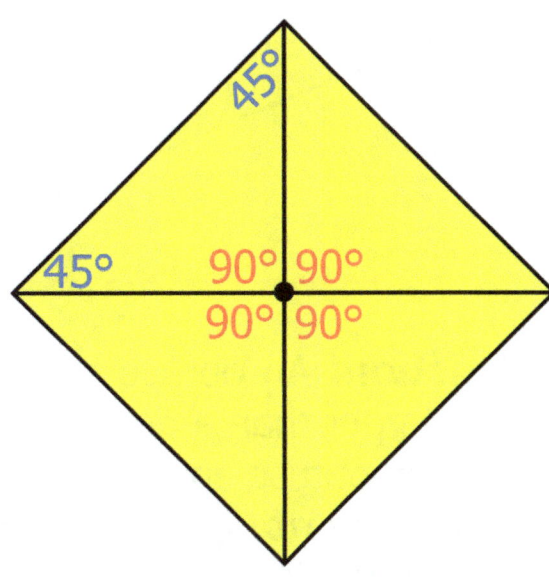

Instruction: If you need to find the measure of the two **base angles** of one of the central angles, subtract the measure of the central angle from 180 (the measure of a triangle's interior angles) and divide the answer by 2.

 $180 - 90 = 90$
 $90 \div 2 = 45°$

Triangles

right

Instruction: A right triangle has one right angle. A right angle measures 90°. An angle marked with a little square indicates that it is a 90° angle.

acute

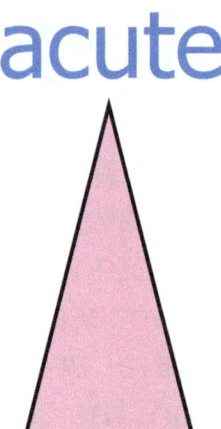

Instruction: An acute triangle has three acute angles. (The three angles are all smaller than a right angle.)

obtuse

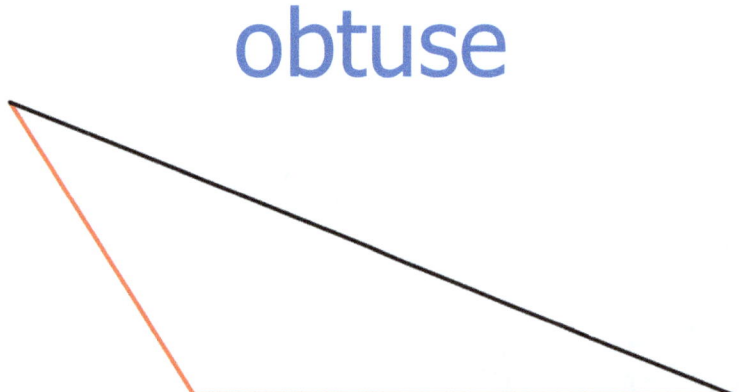

Instruction: An obtuse triangle has one obtuse angle (one of them is larger than a right angle). In the triangle above, the red angle is an obtuse angle. Also, the obtuse and acute triangles on this page are *oblique triangles*. Any triangle that is not right is an oblique triangle.

Triangles

equilateral and equiangular

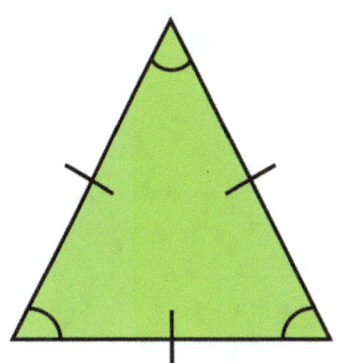

Instruction: The sum of all the interior angles of any triangle is 180°. All sides of an *equilateral* triangle have the same length, and the measure of each interior angle is 60° because 180 ÷ 3 = 60. Since the angles measure the same, it is *equiangular*. Actually, if all the sides of a polygon are equal, then its angles must also be equal. In the same way, if all the angles of a polygon are equal, then its sides must also be equal.

isosceles

Instruction: At least two sides of an isosceles triangle have the same length.

scalene

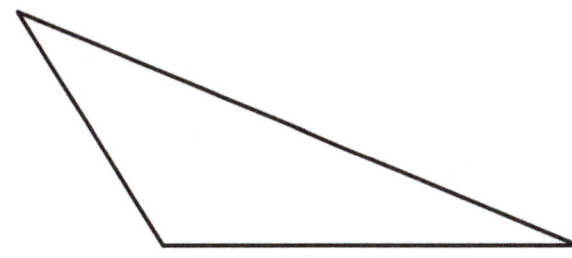

Instruction: No two sides of a scalene triangle have the same length.

Instruction: If you add together what all three interior angles of a **triangle** measure, the sum will always be

180°.

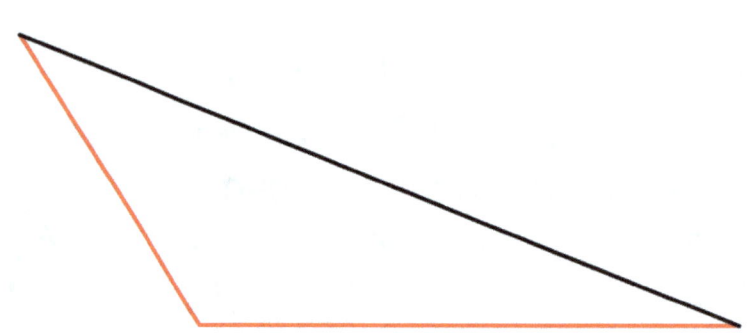

Instruction: If you add together what all four interior angles of a convex **quadrilateral** measure, the sum will always be

360°.

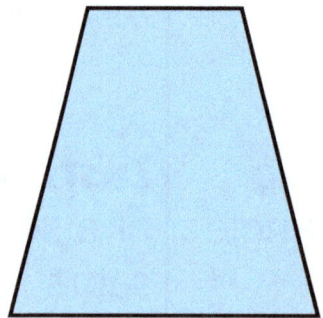

Problems You May See on Your Homework

The measurement of one angle of a triangle is 50° and a second is 75°. What does the third angle measure? Record and solve an equation to find the unknown measurement of the triangle's third angle.

Solution:
$50 + 75 + x = 180$
$125 + x = 180$
$180 - 125 = 55, x = 55$

The measurement of one angle of a trapezoid is 70°, a second is 85°, and a third is 110°. What does the fourth angle measure? Record and solve an equation to find the unknown measurement of the trapezoid's fourth angle.

Solution:
$70 + 85 + 110 + x = 360$
$265 + x = 360$
$360 - 265 = 95, x = 95$

If one angle of a triangle measures 33°, a second angle measures $y°$, and the third angle measures $(2y)°$, what is the value of y?

$y =$ 49°

(**Solution:** $33 + y + 2y = 180$, $33 + 3y = 180$, $3y = 147$, $y =$ **49°**)

If the three angles of a certain triangle measure $x°$, $4x°$, and $5x°$, what is the degree measure of the smallest angle ($x°$) and of the largest angle ($5x°$)?

Smallest angle: 18°

(**Solution:** $x + 4x + 5x = 180$; $10x = 180$; $180 \div 10 = 18$, $x =$ **18°**)

Largest angle: 90°

(**Solution:** $18 \times 5 = 90$)
Since the largest angle ($5x°$) is 5 times as large as the smallest angle, the largest angle is 90°.

Find the degree measure of ∠A. Notice that △ABC is a right triangle. (You will first need to find the value of x.)

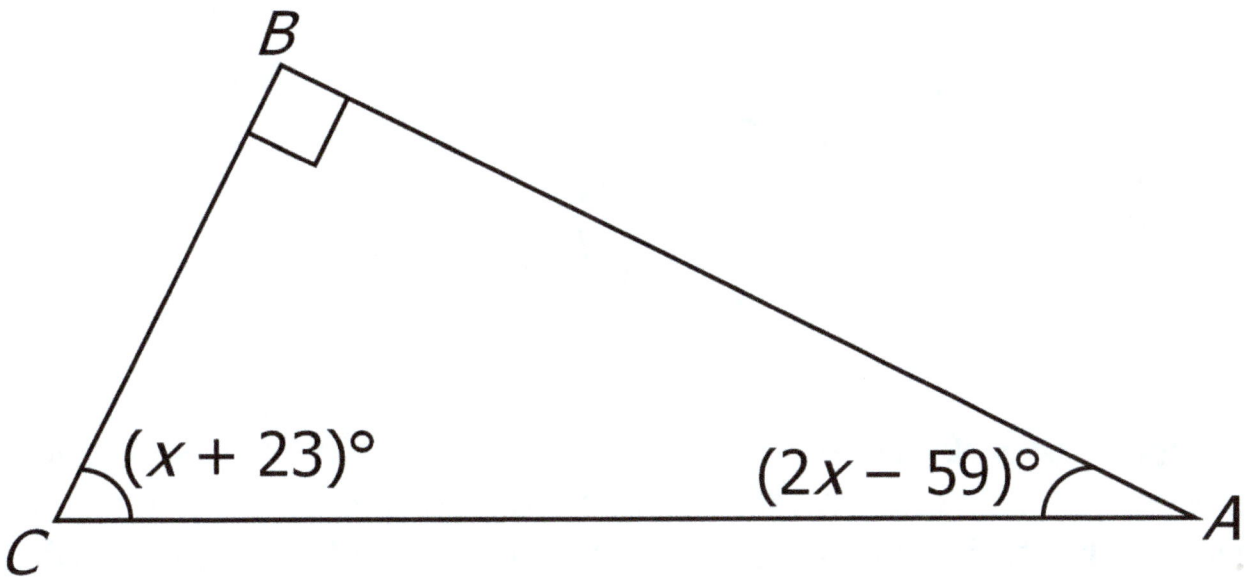

x = 42°

∠A = 25°

Solution:
90 + (x + 23) + (2x − 59) = 180
90 + 23 − 59 = **54**; x + 2x = **3x**
54 + 3x = 180 (180 − 54 = **126**)
3x = 126 (126 ÷ 3 = 42)
x = **42°**

To find the value of angle A, replace the value of x in (2x − 59) with the value you just found and solve.
 ∠A = 2(42) − 59 = 25°, ∠A = **25°**

95

In the triangles below, compare:
1. $a - n$
2. $q - b$

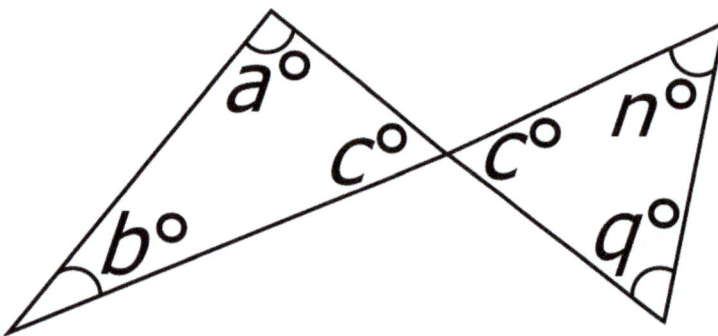

Instruction: Notice that the variable c is used for the vertical angles. (The same variable is used because vertical angles are always the same measure.) Since the three interior angles of a triangle add up to 180°, we know that
($a + b + c = 180$) and that
($n + q + c = 180$).
Thus, we can record
$a + b + c = n + q + c.$
When we subtract c from both sides, we are left with the equation
$a + b = n + q.$

The problem above asks us to compare $a - n$ and $q - b$, so we will need to rearrange the equation $a + b = n + q$ to correspond with the problem.

First, move n to the opposite side of the equation by subtracting it from a. Then, move b to the opposite side of the equation by subtracting it from q. Now we are left with the equation $a - n = q - b$, which shows that 1 and 2 are equal.

Find the value of *n*, *x*, and *y*.

n = 85

x = 31

y = 12

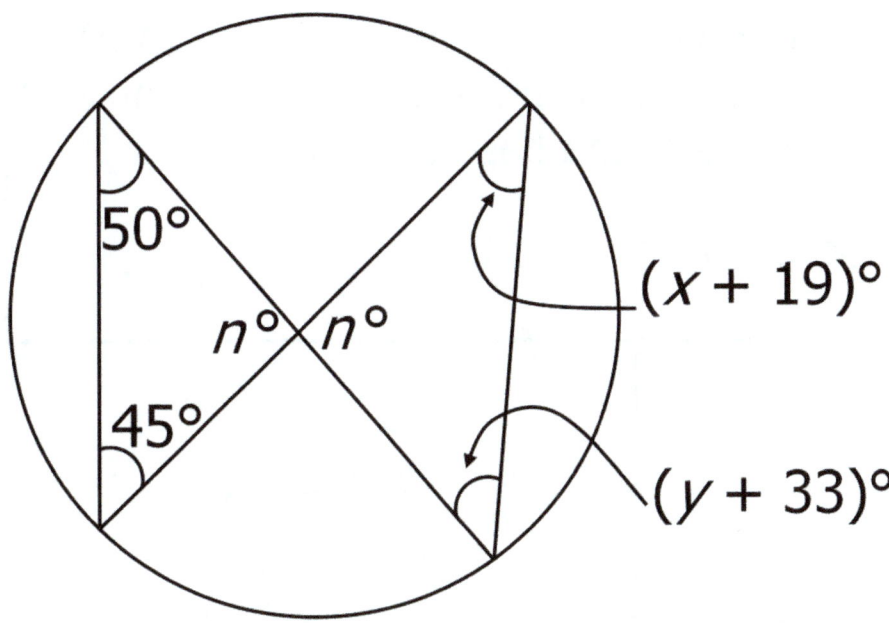

Solution:
50 + 45 + *n* = 180
50 + 45 = 95, 180 − 95 = 85
n = **85**

50 − 19 = 31
x = **31**

45 − 33 = 12
y = **12**

Instruction: The two congruent sides of the isosceles triangle are called legs. The other side is the base. In this triangle, *A* is the **vertex angle**, and *B* and *C* are the **base angles**. The two base angles in an isosceles triangle are congruent. In the trapezoid, the parallel sides are the bases, and the two other sides are the legs.

Problem: In an isosceles triangle, the two angles that do not join the matching sides are congruent. Thus, if the vertex angle of an isosceles triangle measures 34°, then each of its base angles measure _____. You can record an equation and solve it to find the measure of the base angles.

(**Solution:** 180 − 34 = 146, 146 ÷ 2 = 73°)

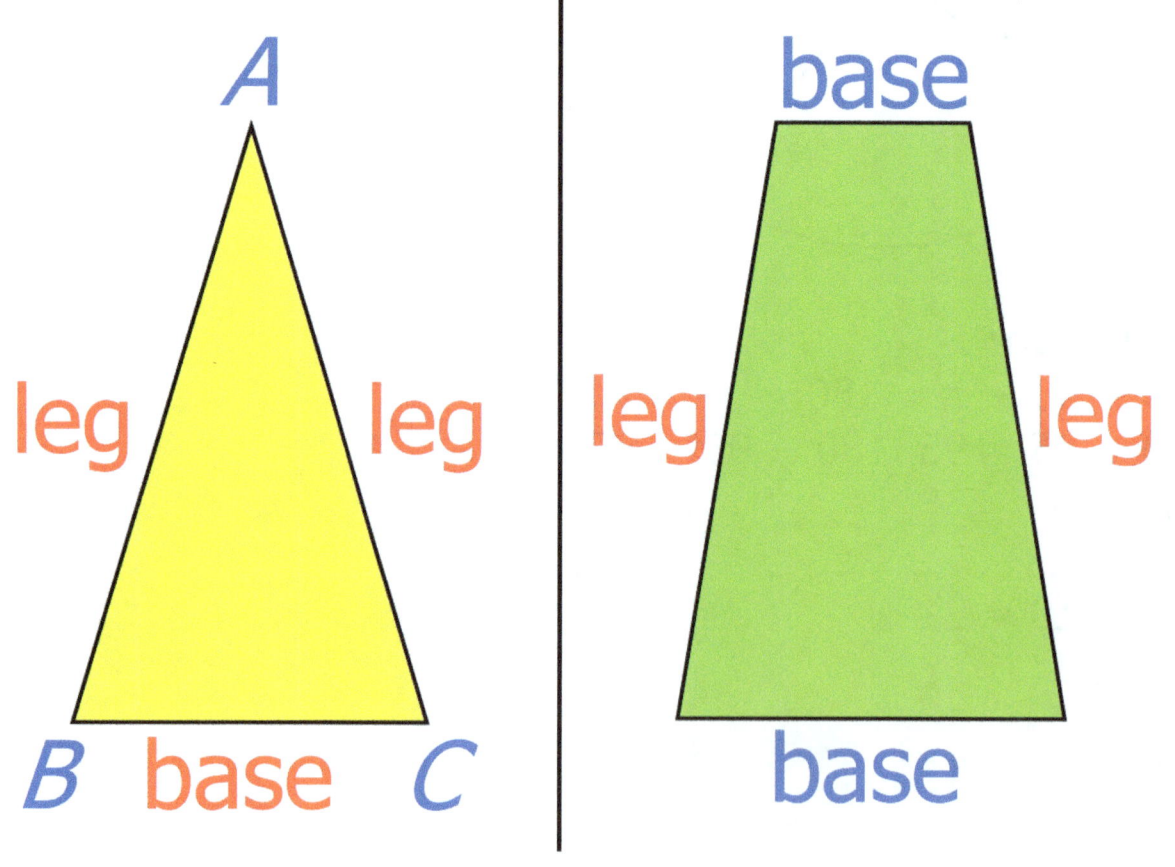

Right Triangle

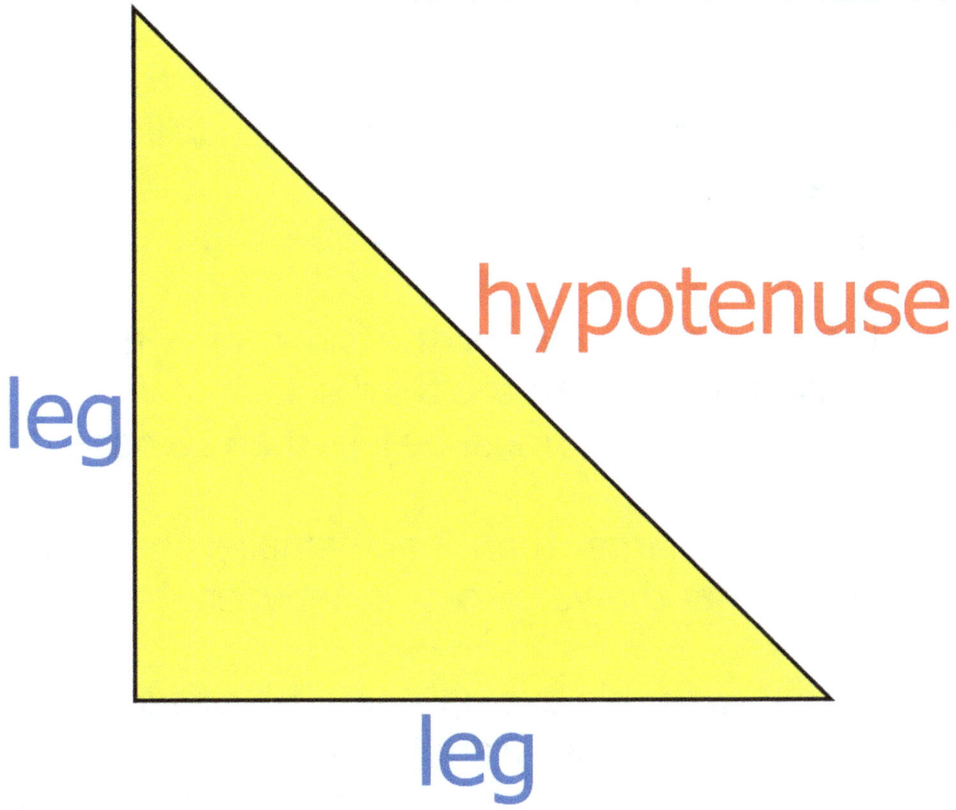

leg

hypotenuse

leg

Instruction: A right triangle has two legs (point to the legs) that form the right angle and a hypotenuse (point to the hypotenuse) across from the right angle.

Angle measurement problem: Record the measurement of one of the acute angles (any reasonable estimate will work) and let students determine the measurement of the other acute angle.

Solution: If one of the acute angles measures 40° and we know that the right angle of the right triangle measures 90°, then the last angle measures 50 degrees (40 + 90 + __ = 180; 180 − 90 − 40 = 50).

Triangle Inequality Rule

- If the length of one side of a triangle is subtracted from the length of another side of the triangle, this difference will be less than the length of the third side.
  Example: In the triangle shown, x must be more than 3 because $8 - 5 = 3$.
- Also, if you add the lengths of two of the sides of a triangle together, the sum will be more than the third side.
  Example: In the triangle shown, x must be less than 13 because $8 + 5 = 13$.

Thus, the value of x has to be larger than 3 and smaller than 13 ($3 < x < 13$). This tells us that the value of x is between 4 and 12.

You should also know that x, in this example, will be closer to the larger number (12) if the angle across from it is also large (nearer to 180° than 0°). In contrast, x will be nearer to the smaller number (4) if the angle across from it is also small (nearer to 0° than 180°).

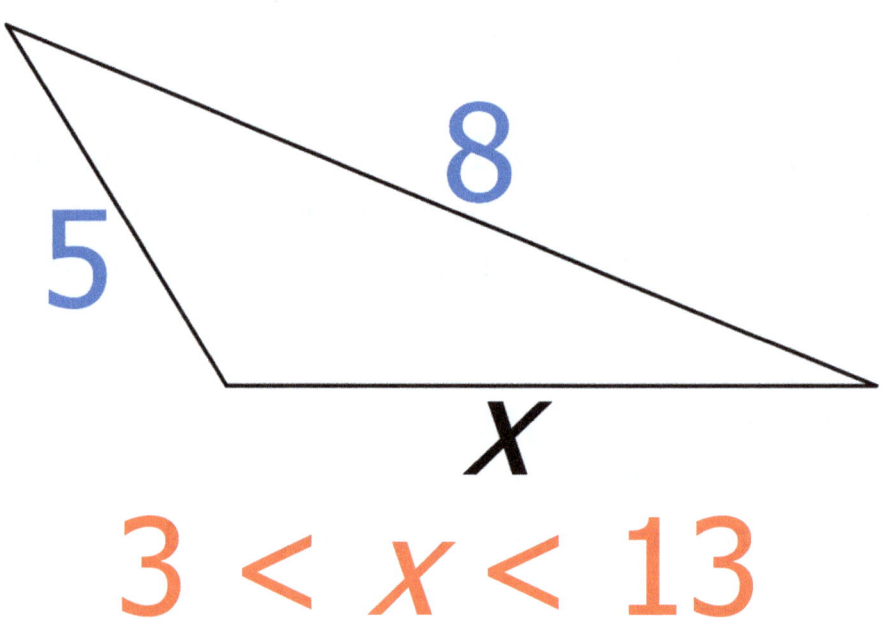

$$3 < x < 13$$

In the figure below, triangle *XYZ* is inscribed in a circle. If the length of side *YZ* is 8, what is the least possible sum of the other two sides of the triangle? Assume that the other two sides are also integers (they are not decimal numbers).

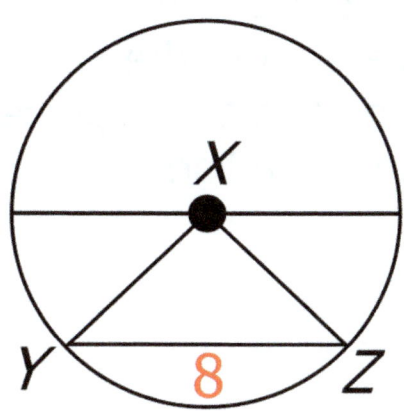

Instruction: According to the "Triangle Inequality Rule" we learned above, the other two sides of the triangle have to add up to more than the third side. Since the third side is 8, we know that the other two sides add up to at least 9. However, because side *XY* and side *XZ* are both *radii* of the circle, they have to be the same length. If they were both 4, then they would equal the length of side *YZ*, so they have to be more than 4. If they are both 5, then together they measure ten (5 + 5 = 10). Thus, 10 would be the least possible sum of the other two sides of the triangle.

Note: You will learn later in this book that a radius is a line segment that extends from the center of a circle to a point on the circle. Also, as noted above, all radii of a circle have the same length.

101

Theorems and Postulates

Instruction: On the next page, you will learn about the Pythagorean Theorem. This is only one of numerous theorems. A **theorem** is a statement or formula that can be proved. Unlike theorems, **postulates** (also called **axioms**) are unproven statements, but postulates are assumed to be true. In fact, postulates are used to help prove the theorems. You have already been learning various postulates and theorems throughout this book, such as the postulate that states "A plane has at least three points that are not on the same line."

Pythagorean Theorem

leg² + leg² = hypotenuse²

Instruction: The Pythagorean Theorem will only work with a right triangle. The theorem was named after Pythagoras—the ancient Greek mathematician who first proved it. The theorem proves that if you square the length of both legs of a right triangle and then add them together, the sum will equal what the hypotenuse equals when it is squared. If the sum does not equal the length of the hypotenuse squared, then it's not a right triangle, even though it might look like one. The formula is $a^2 + b^2 = c^2$ (or leg² + leg² = hypotenuse²).

Four squared is 16, and 3 squared is 9; 16 + 9 = 25. The hypotenuse (5) squared is also 25, so this is a right triangle. (Remember, you square a number by multiplying it by itself.)

$$16 + 9 = 25 \qquad\qquad 5^2 = 25$$

A hypotenuse's length is never negative.

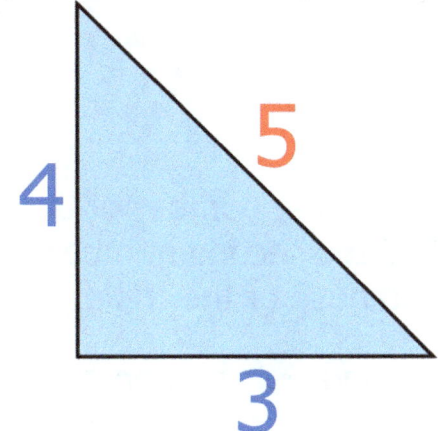

Solution:
$4^2 = 16$, $3^2 = 9$
$16 + 9 = 25$
$5^2 = 25$

Pythagorean Triples

In the right triangle on the previous page, *4*, *3*, and *5* are known as a **Pythagorean triple** because they are positive integers that fulfill the "leg² + leg² = hypotenuse²" equation. To find a Pythagorean triple, use the formula below (*2n* represents one leg of the right triangle; $n^2 - 1$ represents the other leg; and $n^2 + 1$ represents the hypotenuse).

$$2n,\ n^2 - 1,\ n^2 + 1$$

Find the Pythagorean triple if <u>3</u> is the value of *n* (replace *n* with 3). Then verify that the three numbers you identify fulfill the "leg² + leg² = hypotenuse²" equation.

2(<u> 3 </u>) = <u> 6 </u>   <u> 3 </u>² − 1 = <u> 8 </u>   <u> 3 </u>² + 1 = <u> 10 </u>

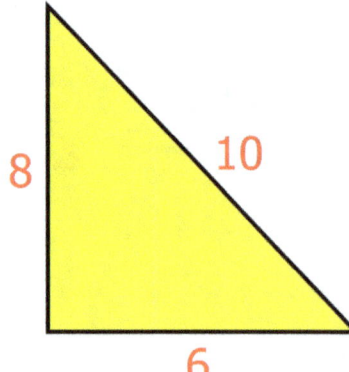

Pythagorean triple: 6, 8, 10

Verification: 6² + 8² = 10², 36 + 64 = 100

Solution: Multiply 2 by the value of *n* to find the first number of the triple; square the value of *n* and then subtract 1 to find the second number; square the value of *n* and then add 1 to find the hypotenuse of the triple.

Teacher: Choose a different value for *n* and let students find the Pythagorean triple.

Problems You May See on Your Homework

Identify the missing measure of the hypotenuse in the right triangle below. Use the radical sign in your answer.

$c = \sqrt{74}$

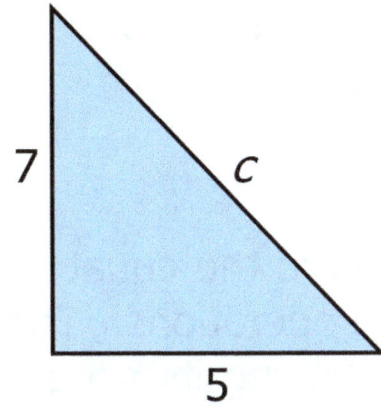

Solution:
$7^2 = 49$, $5^2 = 25$
$49 + 25 = c^2$
$49 + 25 = 74$
$c = \sqrt{74}$

Identify the missing measure of the leg in the right triangle below. (Remember, a hypotenuse's length is never negative.)

$a = \sqrt{144} = 12$

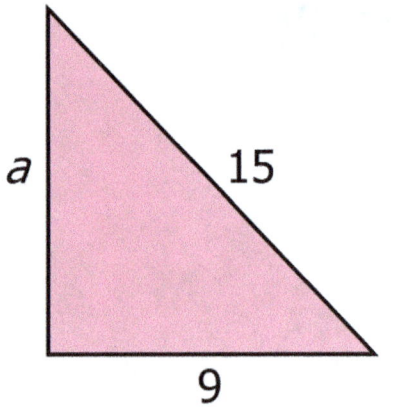

Solution: $a^2 + 9^2 = 15^2$
$a^2 + 81 = 225$
$225 - 81 = 144$, $a = \sqrt{144} =$ **12**

To find the square root of 144 on your calculator, press *shift*, $\sqrt{}$, *144, EXE* (or *=*).

Instruction: To solve problems like this, record the equation and then work backwards using the opposite operations. The opposite operation of squaring a number is finding its square root.

A 17-foot ladder was leaned against a wall 8 feet from the base. How far up the wall does the ladder reach?

Hint: Use the Pythagorean Theorem to help you solve word problems like this. Let the length of the leaning ladder represent the hypotenuse and the other given length represent the measure of the bottom leg.

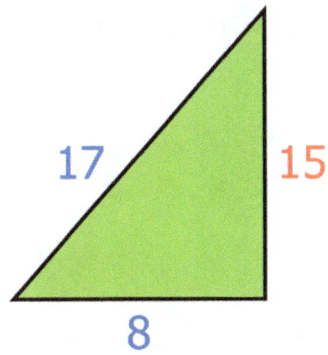

Solution:
$8^2 + b^2 = 17^2$
$64 + b^2 = 289$
$289 - 64 = 225, \sqrt{225} =$ 15 ft

Kailey and Alex left their parents' home at the same time to head back to college. Kailey drove her car due north at a speed of 48 miles per hour, and Alex rode his moped due west at a speed of 14 miles per hour. After traveling for 2 hours, what would the straight-line distance be between the two siblings if they continued at this speed without stopping?

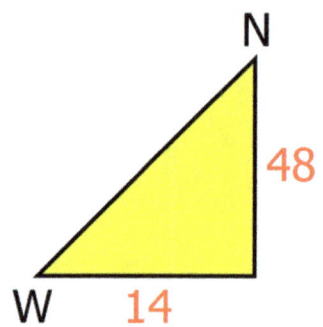

Solution:
$48^2 + 14^2 =$ ___
$2304 + 196 = 2500$, $\sqrt{2500} = 50$
50×2 hours = 100 miles

What is the length of diagonal *AC*?

Hint: One diagonal will divide a rectangle into two congruent right triangles. If you know the length of the sides of the rectangle, which would also be the legs of the two congruent triangles, then you can use the Pythagorean Theorem to find the length of the diagonal, which would also be the hypotenuse of the triangle.

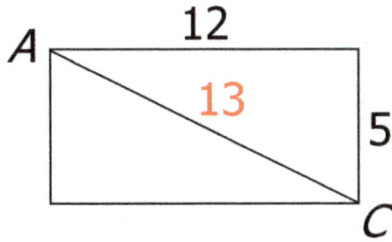

Solution:
$5^2 + 12^2 = c^2$,
$25 + 144 = 169$, $\sqrt{169} = $ **13**

Find the value of $x^2 + y^2$.

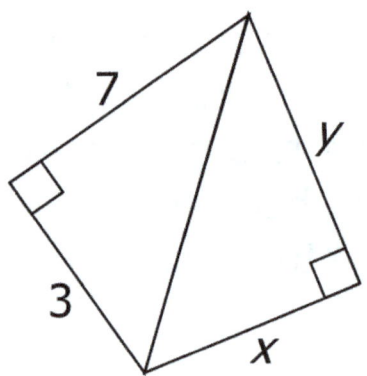

Solution:
$3^2 = 9$, $7^2 = 49$,
$9 + 49 = 58$

Instruction: We don't have to know the precise value of x and of y to answer this question. Instead, we simply need to know that, according to the Pythagorean theorem, the hypotenuse squared $= 3^2 + 7^2$. Notice from the illustration directly above that the two right triangles share the same hypotenuse. Therefore, $x^2 + y^2 = 58$ because $3^2 + 7^2 = 58$.

Find the length of $\overline{AB}$. Notice that point *a* has the coordinates (−4, −3), while point *b* has the coordinates (4, 3).

Hint: Form a right triangle and use the Pythagorean Theorem to find the length of the hypotenuse, which would also be the length of $\overline{AB}$.

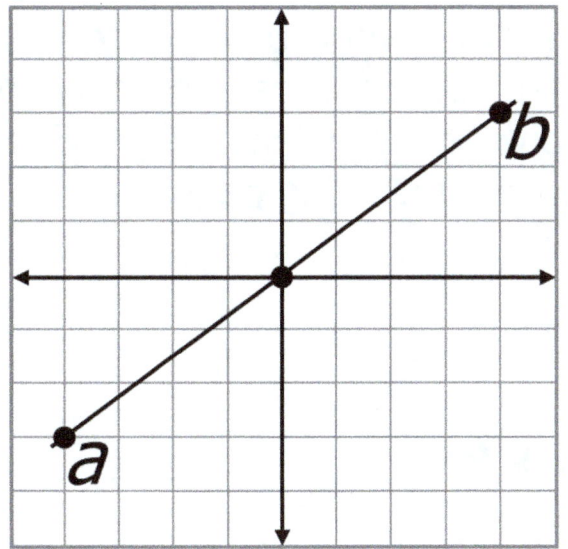
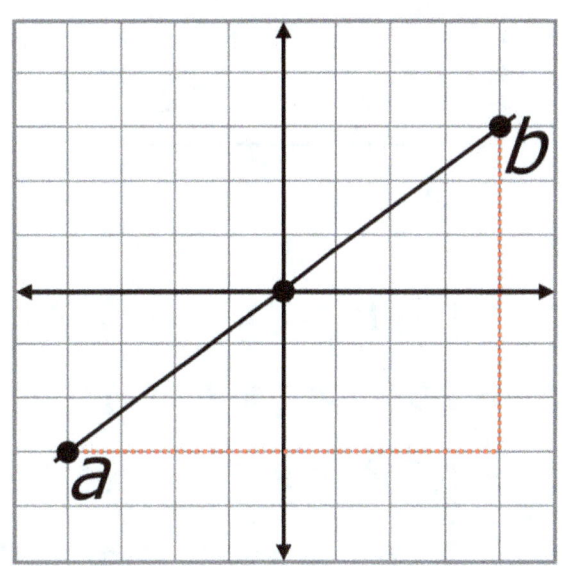

Solution: $6^2 + 8^2 =$ ___
$36 + 64 = 100$
$\sqrt{100} =$ 10

Solution: When you form a right triangle using $\overline{AB}$ as the hypotenuse, the vertical leg is 6 units long, and the horizontal leg is 8 units long. Since $6^2 + 8^2 = 100$ and $\sqrt{100} = 10$, the answer is 10.

If one leg of a right triangle measures x, the other leg measures $5y$, and the hypotenuse measures $x + 3y$, what is the value of x in terms of y?

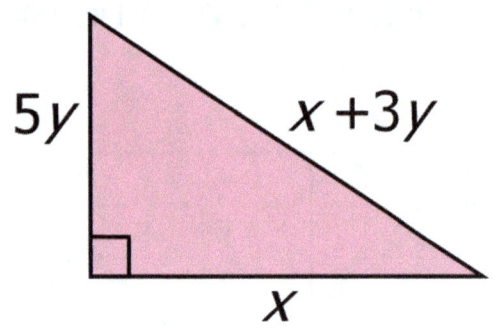

Note: If a diagram like that shown was not provided, you could use the given information to draw your own. Step-by-step instructions on how to solve this problem are provided on the following page.

$x^2 + (5y)^2 = (x + 3y)^2$

$x^2 + 25y^2 = (x + 3y)^2$

$\quad (x + 3y)(x + 3y) = x^2 + 6xy + 9y^2$

$x^2 + 25y^2 = x^2 + 6xy + 9y^2$

$25y^2 = 6xy + 9y^2$

$16y^2 = 6xy$

$16y = 6x$

$x = \dfrac{16y}{6}$ or $\dfrac{8y}{3}$

Solution
Use the Pythagorean Theorem (leg² + leg² = hypotenuse²) to write the equation.
$$x^2 + (5y)^2 = (x + 3y)^2$$

Since the word problem tells us to find the value of x, we will need to get x by itself on one side of the equation.

In $(5y)^2$, the exponent 2 is outside the parentheses, so square the 5 inside the parentheses to get 25 and the variable y to get y^2. You are left with **$25y^2$**.
$$(5y)^2 = \mathbf{25y^2}$$

Now the equation reads:
$$\mathbf{x^2 + 25y^2 = (x + 3y)^2}.$$

In $(x + 3y)^2$, the exponent 2 is outside the parentheses. Use the acronym FOIL to multiply $(x + 3y)$ by itself. FOIL stands for *first, outer, inner, last*.
$$(x + 3y)(x + 3y)$$

First: Multiply both x's (the first symbols) to get **x^2**. $\qquad$ ($\mathbf{x}$ + 3y)($\mathbf{x}$ + 3y)
Outer: Multiply x and $3y$ (the outer symbols) to get **$3xy$**. $\qquad$ ($\mathbf{x}$ + 3y)(x + $\mathbf{3y}$)
Inner: Multiply $3y$ and x (the inner symbols) to get **$3xy$**. $\qquad$ (x + $\mathbf{3y}$)($\mathbf{x}$ + 3y)
Last: Multiply $3y$ by $3y$ (the last symbols) to get **$9y^2$**. $\qquad$ (x + $\mathbf{3y}$)(x + $\mathbf{3y}$)
$$(x + 3y)(x + 3y) = x^2 + 3xy + 3xy + 9y^2 = \mathbf{x^2 + 6xy + 9y^2}$$

Now the equation reads:
$$\mathbf{x^2 + 25y^2 = x^2 + 6xy + 9y^2}.$$

Since there is an x^2 on both sides of the equal sign, cancel them out.
$$\mathbf{25y^2 = 6xy + 9y^2}$$

Since $25y^2$ and $9y^2$ have the same variable (y) and the same exponent (2), you can subtract the $9y^2$ from $25y^2$. Now the equation reads:
$$\mathbf{16y^2 = 6xy}.$$

Get rid of the y on the right side of the equation by dividing it by the y with the exponent 2 on the other side of the equation. This will also get rid of the exponent 2 next to the y. (Hint: Think of y as 1; $1^2 \div 1 = 1$, so we are left with one y.) Now the equation reads:
$$\mathbf{16y = 6x}.$$

Finally, get x by itself on one side of the equation (solve for x in terms of y) by dividing $16y$ by 6.
$$\frac{16y}{6} \text{ or } \frac{8y}{3} = x$$

The lengths of the sides of a certain triangle are given below. Determine if it is a right triangle, an acute triangle, or an obtuse triangle.

2 m, 4 m, 3 m

Instruction: If you know the length of each side of a triangle, you can determine if it is an acute, an obtuse, or a right triangle. Square the two shorter sides and add them together. Then compare the sum with the longer side squared.

Right: If they are equal, the triangle is right.
Obtuse: If the longer side is a greater number, it is an obtuse triangle.
Acute: If the longer side is a lesser number, it is an acute triangle.

$2^2 + 3^2 \bigcirc 4^2$ $4 + 9 \bigcirc 16$

13 < 16

The triangle is an obtuse triangle. When you square the two shorter sides and add them together, the sum is less than the longer side squared.

Two Special Right Triangles

45°-45°-90° Triangle

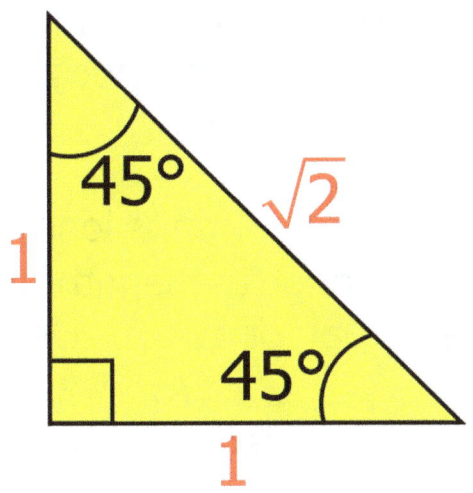

30°-60°-90° Triangle

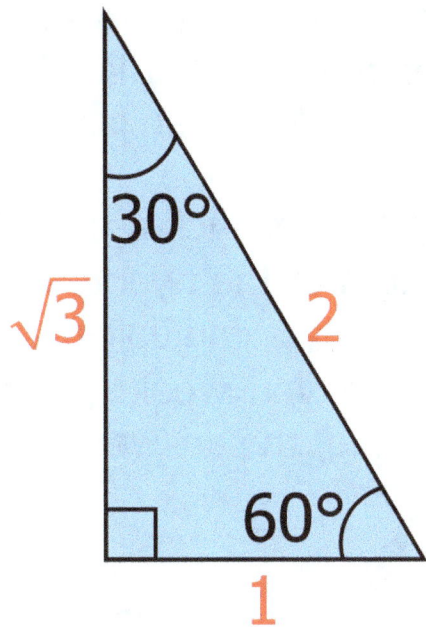

Instruction: The ratio of the shorter leg to the hypotenuse to the longer leg of this special right triangle is $1 : 2 : \sqrt{3}$.

45°-45° Right Triangle

Instruction: An isosceles right triangle (also called a **45°-45° right triangle**) has two acute angles that are the same measure (45°) and two legs that are the same length. Recall that at least two of the sides of an isosceles triangle are the same length, and two angles of a triangle are congruent if the sides across from them are congruent. That is to say, the two angles that do not join the matching sides of an isosceles triangle are congruent.

If both legs of such a triangle are, for example, 4 units long, then the length of the hypotenuse is $4\sqrt{2}$ units long (the length of one of the legs is multiplied by the square root of two). On your calculator, you would press *4, shift, √ , 2, EXE* (or =). After multiplying, the answer can be rounded to the nearest tenth.

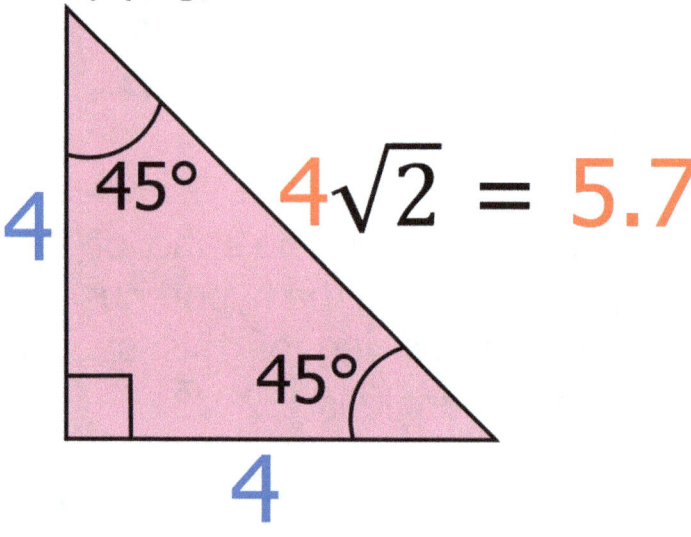

In a 45°-45°-90° right triangle, you find the length of the hypotenuse by multiplying the length of one of its legs by the square root of 2. However, the Pythagorean theorem could also be used to find the length of the side: $4^2 + 4^2 = c^2$, $16 + 16 = 32$, $\sqrt{32} = 5.7$.

116

30°-60° Right Triangle

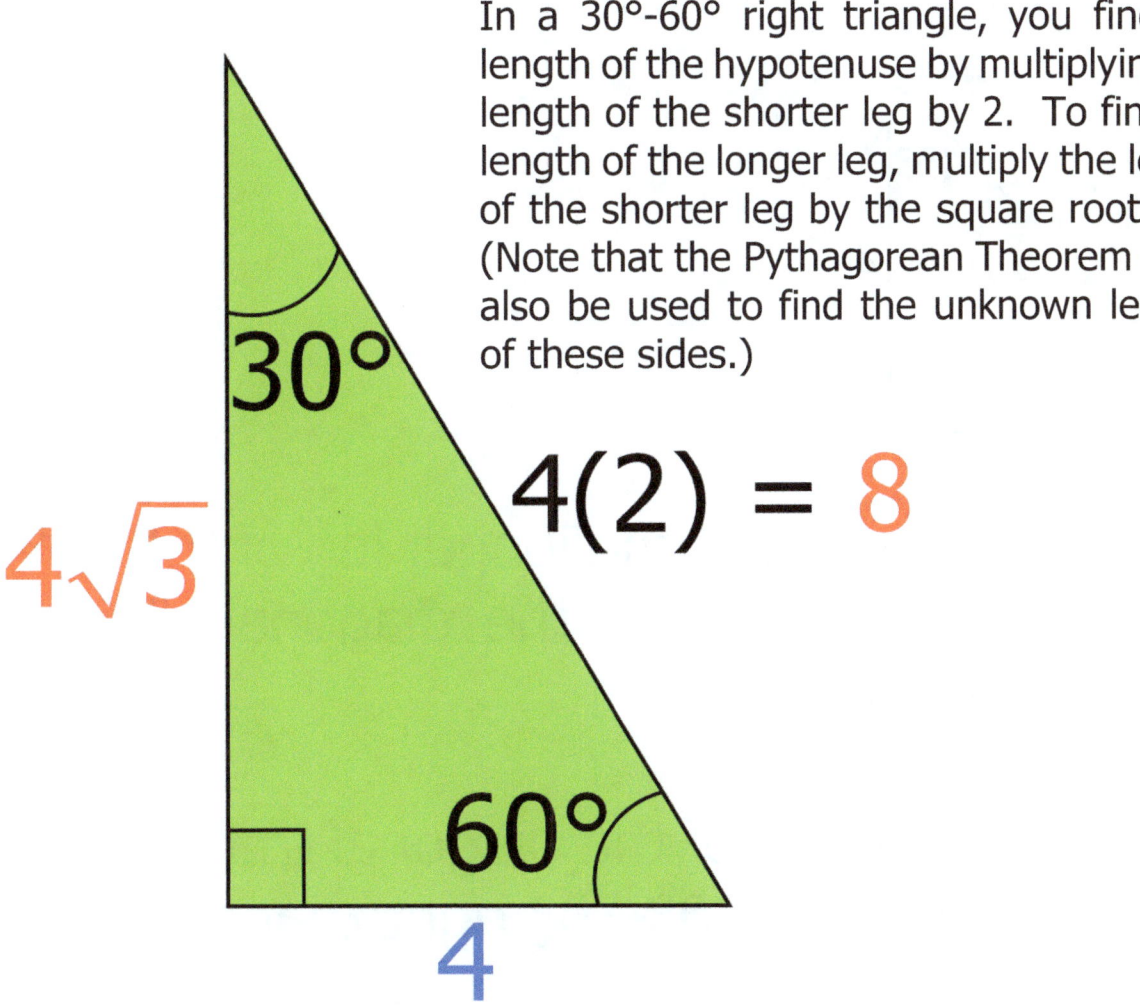

In a 30°-60° right triangle, you find the length of the hypotenuse by multiplying the length of the shorter leg by 2. To find the length of the longer leg, multiply the length of the shorter leg by the square root of 3. (Note that the Pythagorean Theorem could also be used to find the unknown lengths of these sides.)

Instruction: When you divide an equilateral triangle in half, you make two 30°-60° right triangles. If you know the length of the shorter leg of a **30°-60° right triangle**, you can multiply its length by 2 to find the length of its hypotenuse. Also, if the length of the shorter leg is, for example, 4 units long, then the length of the longer leg is $4\sqrt{3}$ units long (the length of the shorter leg is multiplied by the square root of three). If you knew the length of the hypotenuse rather than the shorter leg, how could you find the length of the shorter leg? (*You would divide the hypotenuse by 2.*)

117

Identify the length of the diagonal that cut the square into two 45°-45° right triangles.
(Recall that one diagonal will divide a square into two 45°-45° right triangles and that the diagonal would be the hypotenuse.)

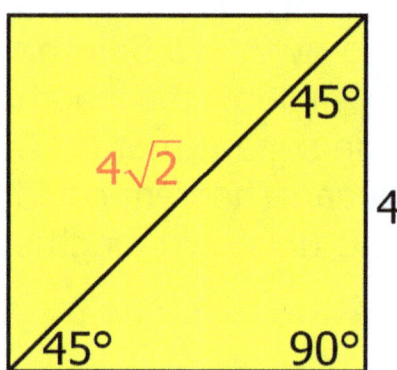

Identify the length of the longer leg of the 30°-60° right triangles that the diagonal divided the rectangle into.

Identify the length of the diagonal that cut the rectangle into two 30°-60° right triangles.

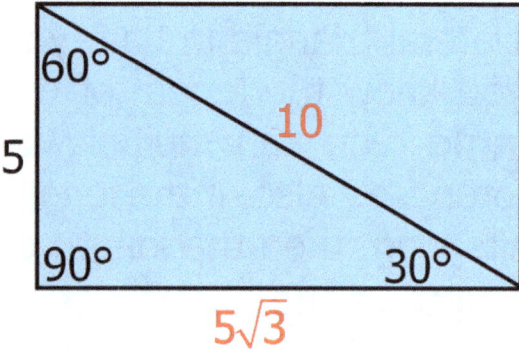

If *XYZ* is an equilateral triangle and the length of the perpendicular bisector $\overline{XW}$ is 3, then the length of $\overline{XZ}$ is what?

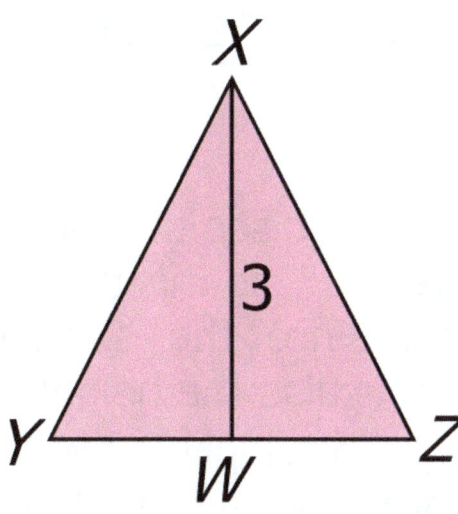

$$\overline{XZ} = \frac{6}{\sqrt{3}} = 2\sqrt{3}$$

$$\frac{\sqrt{3}}{2} = \frac{3}{XZ}$$

$$3 \times 2 = 6$$

$$6 \div \sqrt{3} = \frac{6}{\sqrt{3}} = 2\sqrt{3}$$

Instruction: Recall that a perpendicular bisector cuts an equilateral triangle into two congruent 30°-60° right triangles. The ratio of the shorter leg to the longer leg to the hypotenuse of such a triangle is $1:\sqrt{3}:2$. We can set up a proportion using this information. In this particular problem, we know what the longer leg equals (3), and we need to find what the hypotenuse equals. Since the ratio of the longer leg to the hypotenuse of any 30°-60°-90° triangle is $\sqrt{3}:2$, we record $\sqrt{3}$ over 2 in the proportion. In the triangle above, the length of the longer leg is 3, so 3 is recorded across from the $\sqrt{3}$.

$$\frac{\sqrt{3}}{2} = \frac{3}{XZ}$$

Now cross multiply 3 and 2 and divide the product by $\sqrt{3}$.

$$3 \times 2 = 6, \text{ and } 6 \div \sqrt{3} = \frac{6}{\sqrt{3}} = 2\sqrt{3}$$

The steps below were used to simplify $\frac{6}{\sqrt{3}}$ to $2\sqrt{3}$.

Since $\sqrt{3}$ is the denominator, multiply $\frac{6}{\sqrt{3}}$ by $\frac{\sqrt{3}}{\sqrt{3}}$. $\frac{6}{\sqrt{3}} \times \frac{\sqrt{3}}{\sqrt{3}} = \frac{6\sqrt{3}}{3}$

The six in $\frac{6\sqrt{3}}{3}$ can be divided evenly by the denominator 3, so change 6 to 2 (because $6 \div 3 = 2$). This cancels out the 3 in the denominator so that $\frac{6}{\sqrt{3}}$ is simplified to $2\sqrt{3}$.

A young girl cast a 2-foot shadow at 12:00 noon. How tall is the girl if there is a 60° angle from the end of the shadow to the tip of her head? Round your answer to the nearest tenth.

Instruction: From the illustration, you can see that one angle measures 90° and the other measures 60°, so this has to be a 30°-60° right triangle. Therefore, since the shorter leg measures 2 ft., the longer leg (which is also the height of the girl), has to measure $2\sqrt{3}$. When you multiply 2 by $\sqrt{3}$ and round the decimal to the nearest tenth, you find that the girl is 3.5 ft. tall.

Solution: $2\sqrt{3} = 3.5$ ft

Finding a triangle's shortest, longest, & second-longest side

- In a triangle, the largest angle will always be on the opposite side of the longest side.
- The second-largest angle will always be on the opposite side of the second-longest side.
- The smallest angle will always be on the opposite side of the shortest side.

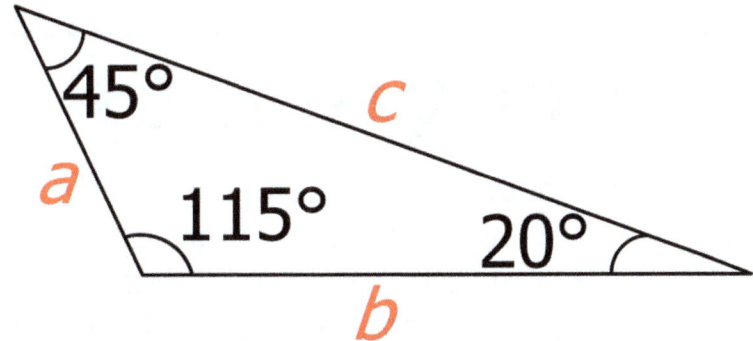

If the length of side *a* is 5 cm and the length of side *c* is 7 cm, then the value of *b* could be what? (Assume that *b* is not a decimal number.)

Answer: The second-largest side could equal 6 cm (the number between 5 and 7).

In the figure above, **A** is the shortest side because it's on the opposite side of the smallest angle (the 20° angle).
B is the second-longest side because it's on the opposite side of the second-largest angle (the 45° angle).
C is the longest side because it's on the opposite side of the largest angle (the 115° angle).

Similar

Instruction: Similar figures have the same shape, but their sizes might differ. Their matching angles have the same measure, and their matching sides are proportional. The symbol "~" shows that the blue triangle is similar to the pink triangle.

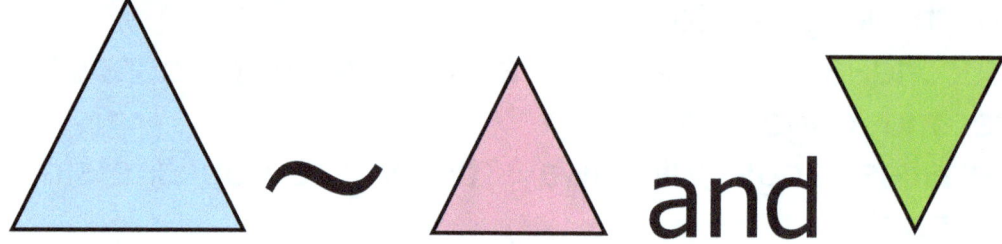

Congruent

Instruction: Congruent shapes are also similar, but they must have the same shape and the same size. The first triangle to the left is congruent to the second triangle because the matching angles of the two triangles have the same measure, and the matching sides have the same length. Angle A is congruent to angle D; $\angle B$ is congruent to $\angle E$, and $\angle C$ is congruent to $\angle F$. Side AB is congruent to side DE, and so forth.

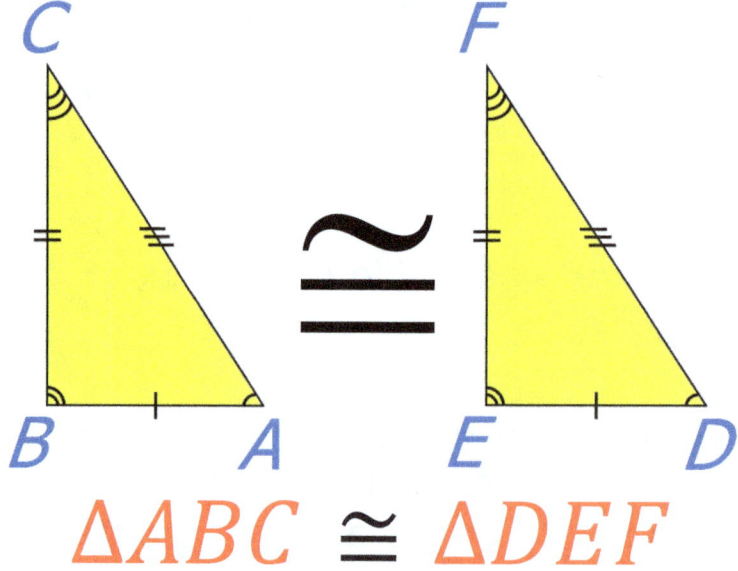

$$\triangle ABC \cong \triangle DEF$$

The notation above reads triangle ABC is congruent to triangle DEF. However, it would be incorrect to record that "$\triangle ABC \cong \triangle EFD$" or "$\triangle ABC \cong \triangle FDE$."

Notes

Congruence Properties

SSS   SAS

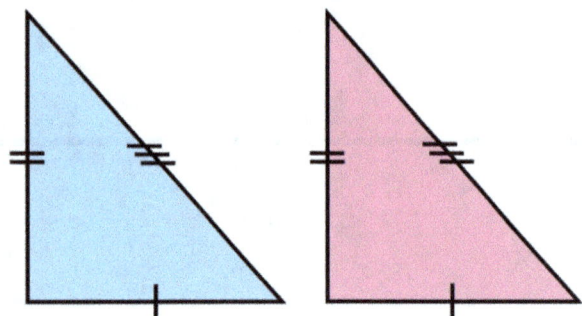
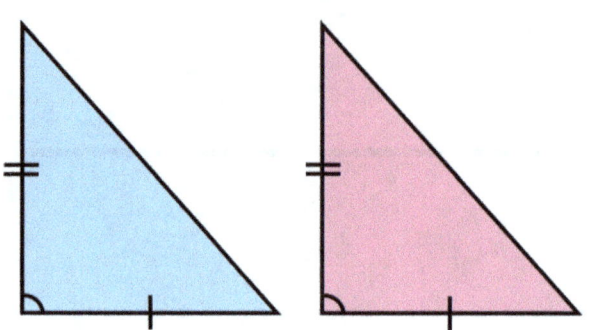

Instruction: You know that two triangles are congruent if each side of one triangle is congruent to the matching side of another triangle. Notice that the matching sides are marked with the same number of lines.

Instruction: You know that two triangles are congruent if the side-angle-side (in that order) of one triangle is congruent to the matching side-angle-side of another triangle. When tracking the order, you can skip over a side but not an angle.

Congruence Properties

SAAA

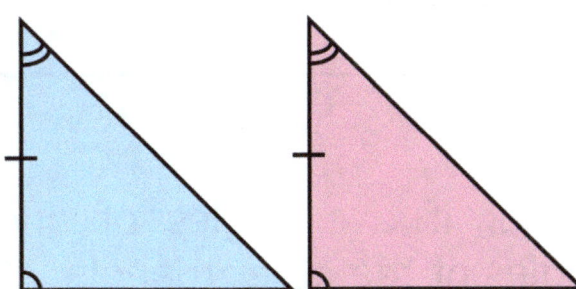

Instruction: You know that two triangles are congruent if the side-angle-angle-angle of one triangle is congruent to the matching side-angle-angle-angle of another triangle. Notice that the matching angles are marked with the same number of arcs. Although the third angle of these triangles is not marked, we know that they have to be congruent since the other two angles of the triangles are congruent.

HL

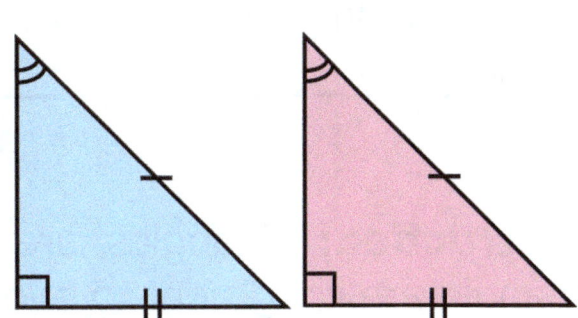

Instruction: You know that two triangles are congruent if the hypotenuse and leg of one triangle are congruent to the hypotenuse and corresponding leg of another triangle.

Similarity Properties

SSS

Instruction: Two triangles are similar if each side of one triangle is in proportion with the matching side of another triangle.

AA

Instruction: Triangles are similar if two angles of one triangle are congruent to two angles of the other triangle.

SAS

Instruction: If two sides of one triangle are in proportion with the two matching sides of a second triangle and the angles between the two sides are congruent, then the triangles are similar. (Recall that *SSS* and *SAS* are also two of the congruence properties.)

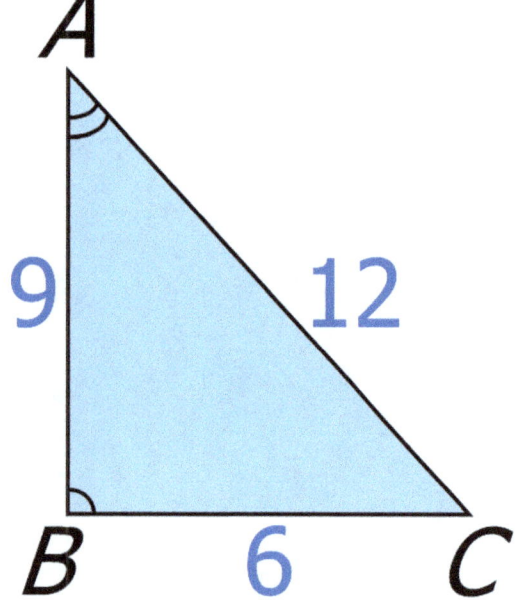

The next page shows that all three sides of the triangles on this page are in proportion.

126

Similar Polygons

Instruction: In similar polygons, the matching angles are congruent, and the matching sides are in proportion. In the figure below, the length of side *AB* is 9, and the length of the matching side *DE* is 3, which can be reduced to 3 over 1. The length of side *BC* is 6, and the length of the matching side *EF* is 2, which can also be reduced to 3 over 1. Finally, the length of side *AC* is 12, and the length of the matching side *DF* is 4, and it, too, can be reduced to 3 over 1.

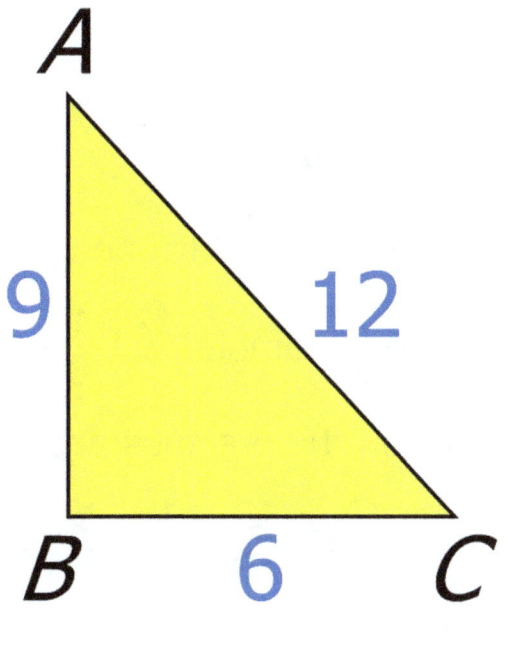

$$\frac{AB}{DE} = \frac{9}{3} \text{ or } \frac{3}{1}$$

Since the lengths of the matching sides have the same ratio, these sides are proportional.

$$\frac{BC}{EF} = \frac{6}{2} \text{ or } \frac{3}{1} \qquad \frac{AC}{DF} = \frac{12}{4} \text{ or } \frac{3}{1}$$

Scale Factors

In the triangles on the previous page, the scale factor from **right to left** (from $\triangle DEF$ to $\triangle ABC$) is $\frac{3}{1}$ or 3 because we can multiply a side from the right triangle by 3 to find the corresponding side of the left triangle ($3 \times \mathbf{3} = 9$, $2 \times \mathbf{3} = 6$, $4 \times \mathbf{3} = 12$). Notice that $\frac{3}{1}$ is the proportion we got on the previous page when we started with the left triangle, not the right: $\frac{9}{3} = 3$, $\frac{6}{2} = 3$, and $\frac{12}{4} = 3$.

$$\overleftarrow{SF} = 3$$

The scale factor from **left to right** (from $\triangle ABC$ to $\triangle DEF$) is $\frac{1}{3}$ because we can multiply a side from the left triangle by $\frac{1}{3}$ to find the corresponding side of the right triangle. For example, $6 \times \frac{1}{3} = 2$. (Notice that $\frac{1}{3}$ is the proportion we would get on the previous page if we started with the right triangle instead of the left: $\frac{3}{9} = \frac{1}{3}$, $\frac{2}{6} = \frac{1}{3}$, and $\frac{4}{12} = \frac{1}{3}$.)

$$\overrightarrow{SF} = \frac{1}{3}$$

Also notice that the arrowhead is on its way to the left when the notation is used to show the scale factor from right to left, while the arrowhead is on its way to the right when the notation is used to show the scale factor from left to right.

Using Proportions to Find Missing Side Lengths of Similar Triangles

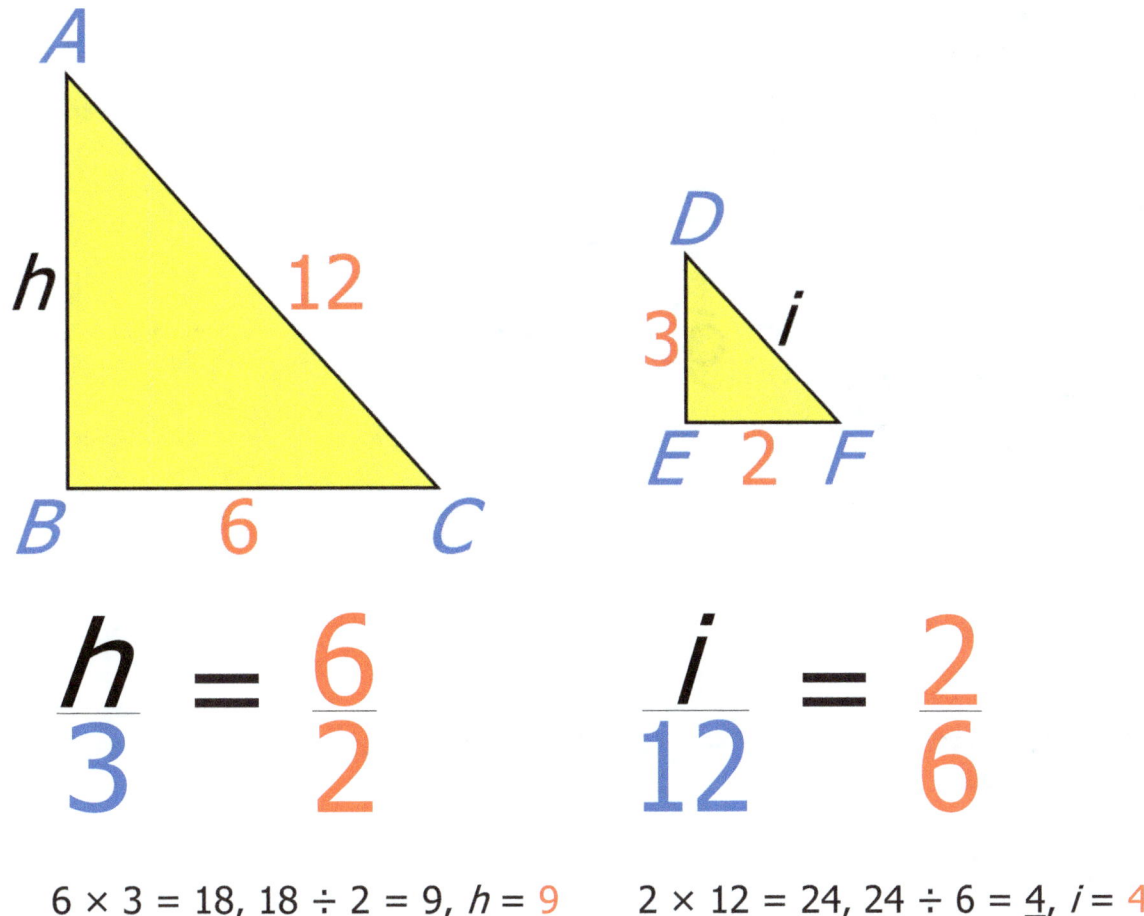

$$\frac{h}{3} = \frac{6}{2} \qquad \frac{i}{12} = \frac{2}{6}$$

6 × 3 = 18, 18 ÷ 2 = 9, h = 9   2 × 12 = 24, 24 ÷ 6 = 4, i = 4

Instruction: The triangles above are similar, so you can use proportions to find the lengths of sides h and i. The length of side AB is unknown, and the length of the matching side DE is 3. Record the ratio h over 3 and record an equal sign after it. You must know both numbers of the second ratio you record, so you must use 6 over 2. (The length of side BC is 6, and the length of the matching side EF is 2.) Notice that 6, which is the length of side BC, is recorded across from the h because it is from the same triangle as h. Multiply the two known cross numbers (6 × 3 = 18) and then divide the product by the number you did not multiply to find that the length of h is nine (18 ÷ 2 = 9).

Now, set up the second ratio to find the length of i (DF over the matching side AC and EF over the matching side BC). Multiply the two known cross numbers (2 × 12 = 24) and divide the product by six (24 ÷ 6 = 4).

If $\overline{CD}$ bisects angle C, what is x.

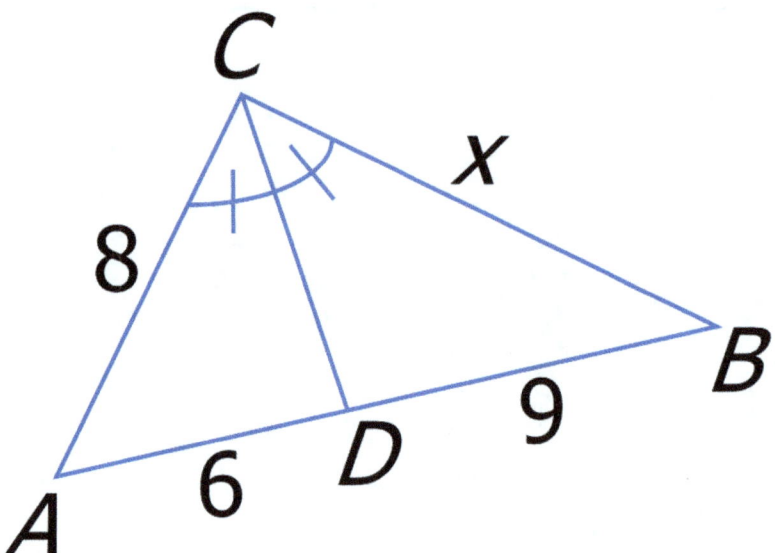

x = 12

Solution: $\frac{8}{6} = \frac{x}{9}$ (8 × 9 = 72, 72 ÷ 6 = 12)

Find *n*.

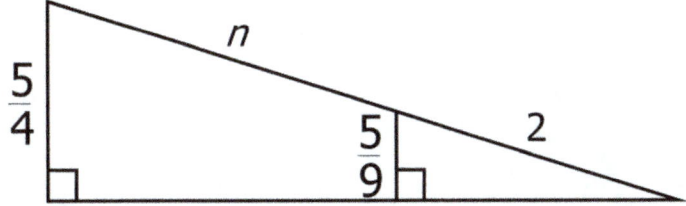

Instruction: The two known corresponding sides of these similar triangles are fractions. When we divide these fractions, we get a quotient of 9/4. Record an equation that shows when the other two corresponding sides (*n* + 2) and 2 are divided, they also equal 9/4. Then we can solve for *n*.

Solution:

$$\left(\frac{5}{4}\right) \div \left(\frac{5}{9}\right) = \frac{5}{4} \times \frac{9}{5} = \frac{45}{20} = \frac{9}{4}$$

$$\frac{9}{4} = \frac{n+2}{2}$$

$9 \times 2 =$ **18**; $4(n + 2) =$ **4*n* + 8**

$18 = 4n + 8$

$10 = 4n$

$\frac{10}{4} = \frac{5}{2}$ or 2.5

$n = \frac{5}{2}$ or 2.5

Find *x* and *z*.

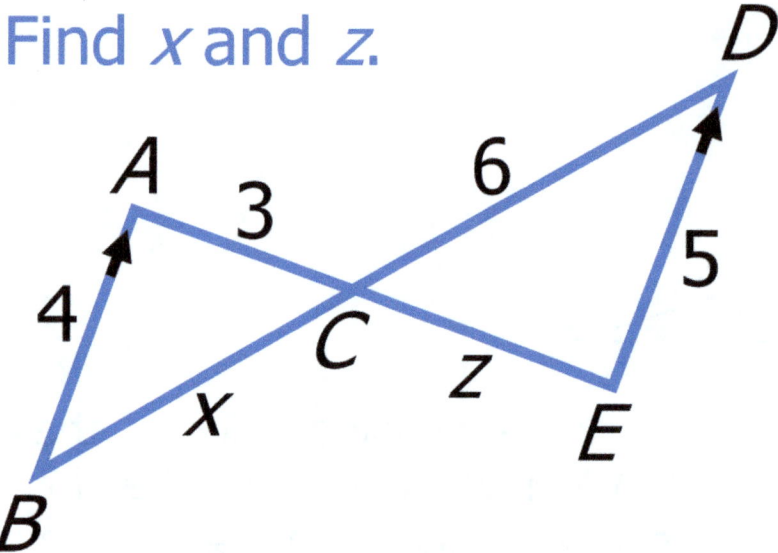

$\frac{4}{5} = \frac{x}{6}$, $4 \times 6 = 24$, $24 \div 5 = \frac{24}{5}$ or 4.8; $x = \frac{24}{5}$ or 4.8

$\frac{4}{5} = \frac{3}{z}$, $3 \times 5 = 15$, $15 \div 4 = \frac{15}{4}$ or 3.75; $z = \frac{15}{4}$ or 3.75

Instruction: There are several things to notice about the figure above. Besides forming two triangles, two of the lines are parallel as indicated by the arrows (side *AB* and side *DE*). Focus upon line *AE*. Notice that it is a transversal line (as is line *BD*) that cuts the two parallel lines. Now recall that when two parallel lines are intersected by a transversal, the alternate interior angles are congruent. This means,

∠*A* ≅ ∠*E*.

We also know that the vertical angles are congruent:

∠*C* ≅ ∠*C*.

Consequently, we know that the two triangles are similar because of the *AA similarity property*, which states that: "Triangles are similar if two angles of one triangle are congruent to two angles of the other triangle." We also know that the matching sides of similar triangles are proportional, so we can use proportions to find the length of sides *x* and *z*.

Are triangles *ABC* and *ADE* similar? How do you know?

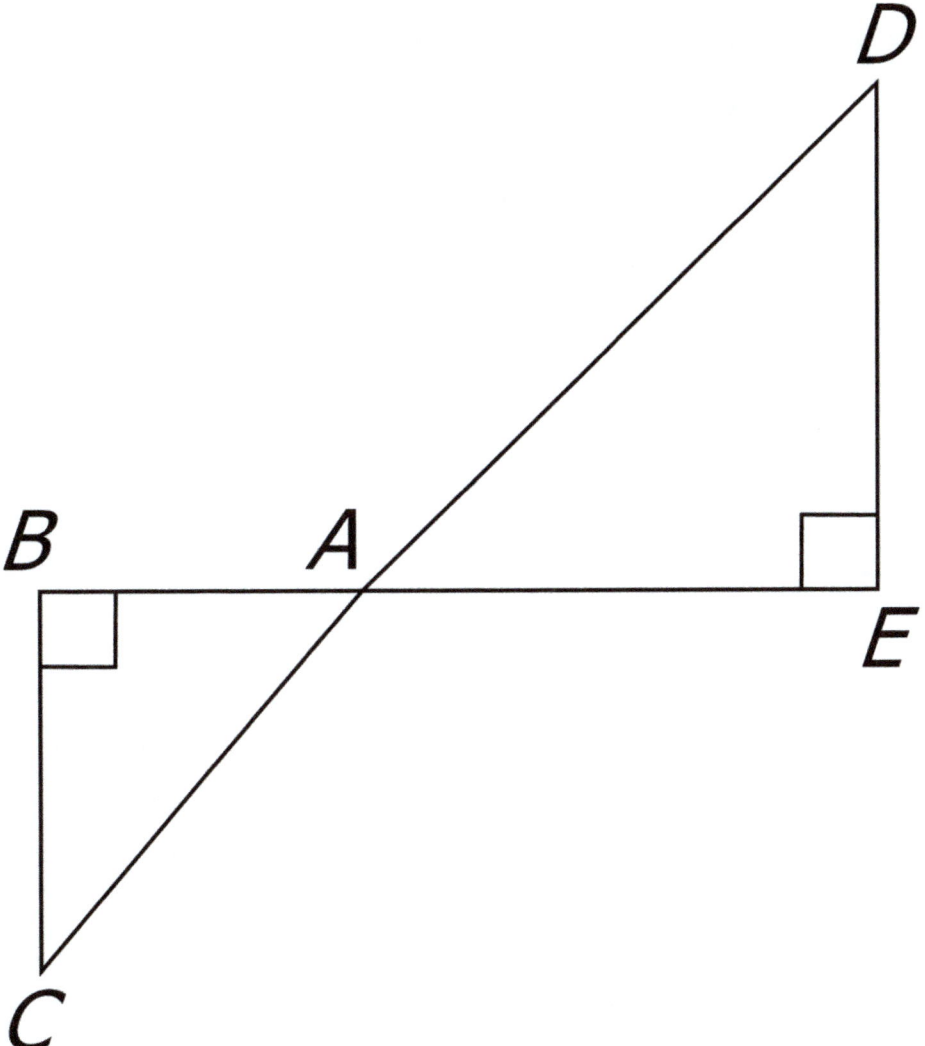

Answer: Yes. We know that two of its angles are the same measure—the 90° angles (*B* and *E*) and the vertical angles (*A*). Since the sum of all the interior angles of a triangle is 180°, angles *C* and *D* must also be the same measure.

If $\overline{CB}$ is parallel to $\overline{ED}$, then triangle *ABC* and triangle *ADE* are ***similar***. How can you be certain that the two line segments are parallel?

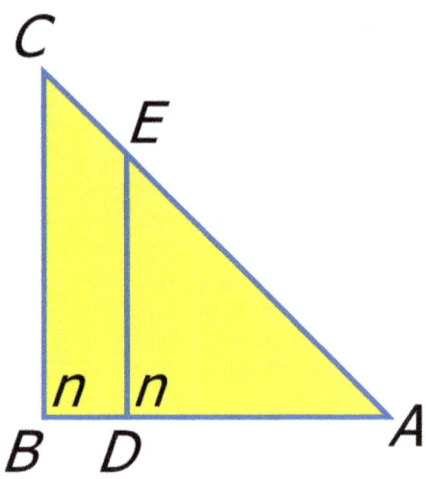

Answer: We know that $\overline{CB}$ and $\overline{ED}$ are parallel because, when they intersect $\overline{BA}$, they both form the same angle (*n*), which looks like a 90° angle in this example but could be slightly more or less.

Find *a*, *b*, and *c*.

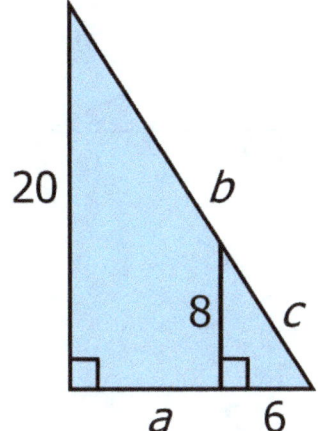

$a = 9$

$b = 15$

$c = 10$

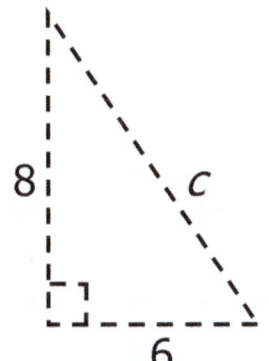

Separate the triangles and use the Pythagorean Theorem to find the value of *c*.

$8^2 + 6^2 = c^2$
$64 + 36 = 100, \sqrt{100} = 10$
c = 10

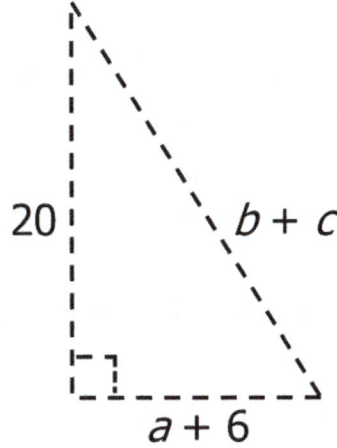

Use proportions to find the value of *a*. Then you can use the Pythagorean Theorem to find the value of *b*.

$\dfrac{20}{8} = \dfrac{a+6}{6}$

$20 \times 6 = \mathbf{120}$
$8(a + 6) = \mathbf{8a + 48}$
$120 = 8a + 48, 72 = 8a, \mathbf{a} = 9$

$a + 6 = 15$
$20^2 + 15^2 =$
$400 + 225 = 625$
$\sqrt{625} = 25$
$25 - c =$
$25 - 10 = 15, \mathbf{b} = 15$

Find x, y, and z.

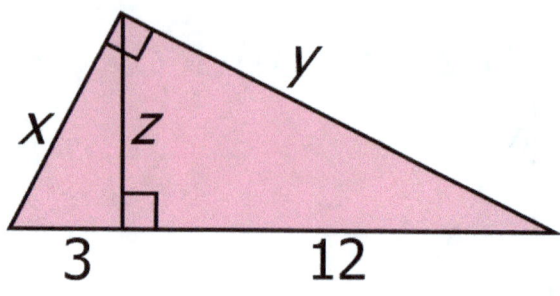

Instruction: The figure to the right consists of three right triangles, so let's begin by illustrating the triangles separately (see below).

$x = \sqrt{45}$

$y = 6\sqrt{5}$

$z = 6$

  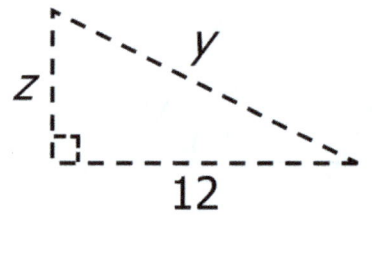

We can use the first two triangles illustrated above to set up a proportion because one leg of the first triangle has the variable x, and we know the number of the corresponding side of the middle triangle (3). Further, the same variable (x) marks the hypotenuse of the middle triangle, and we know that the number of the hypotenuse of the first triangle is 15.

$\frac{x}{3} = \frac{15}{x}$ $15 \times 3 = 45$ $x^2 = 45$ $x = \sqrt{45}$

After finding the value of x, we can use that value and the Pythagorean Theorem in the middle triangle to find the value of z.
$z^2 + 3^2 = (\sqrt{45})^2$ $z^2 + 9 = 45$ $(45 - 9 = 36)$ $z^2 = 36$
$z = \sqrt{36} = 6$

After finding the value of z, we can use that value and the Pythagorean Theorem to find the value of y in the final triangle.
$6^2 + 12^2 = y^2$ $36 + 144 = 180$ $y = \sqrt{180} = 6\sqrt{5}$

Use the largest triangle and the Pythagorean Theorem to **check your answer**.
$y^2 + x^2 = 15^2$ $(6\sqrt{5})^2 + (\sqrt{45})^2 =$ $180 + 45 = 225$ $\sqrt{225} = 15$

A polygon with 10 sides is similar to another polygon. The length of one side of the larger polygon is 4 times greater than the length of the matching side of the other polygon. Record a ratio that compares the area of the bigger polygon to that of the smaller one.

(Hint: You do not need to find the areas of the two polygons to answer this question. Also, although this problem tells us how many sides the similar polygons have, this information is not needed to answer the question.)

$$\left(\frac{4}{1}\right)^2 = \frac{16}{1}$$

A triangle is similar to another triangle. The length of one side of the larger triangle is 3/2 times longer than the length of the matching side of the other triangle. Record a ratio that compares the area of the smaller triangle to that of the larger one.

(Hint: You do not need to find the areas of the two triangles to answer this question.)

$$\left(\frac{3}{2}\right)^2 = \frac{9}{4} = \frac{4}{9}$$

Since the question is asking for the ratio of the area of the smaller triangle to that of the larger one, the answer is 4/9 rather than 9/4.

Median

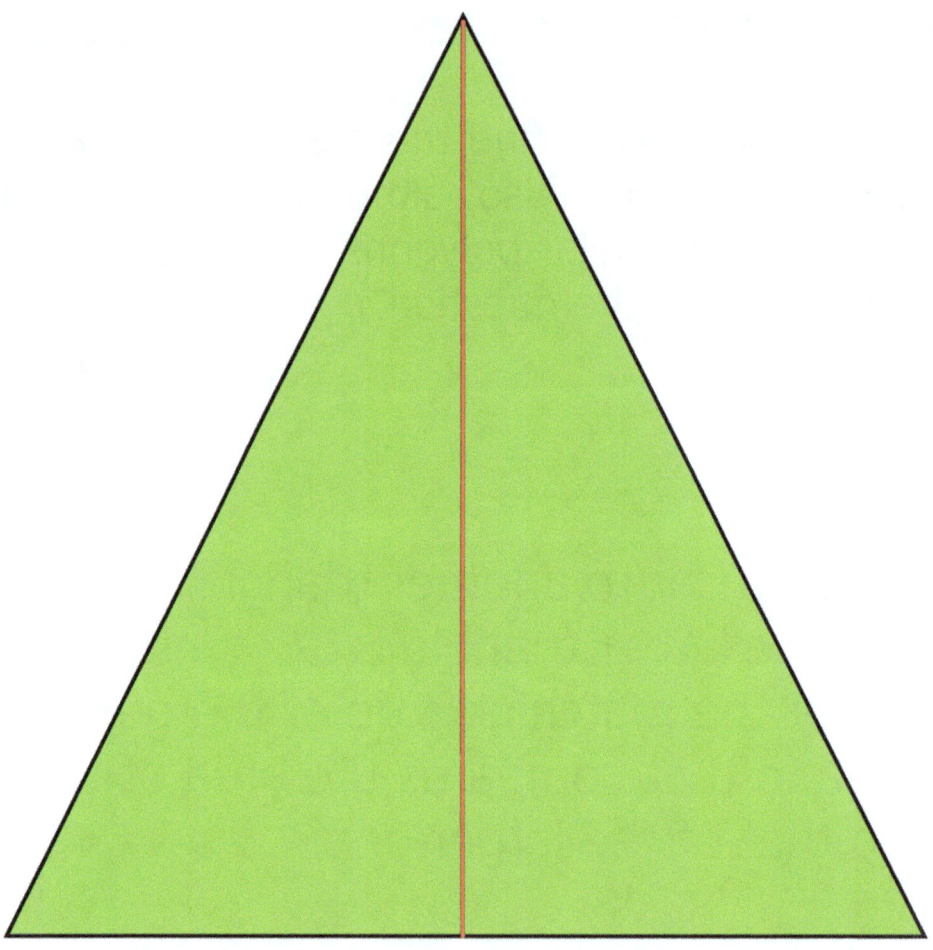

Instruction: In a triangle, a segment that stretches from a vertex to the middle of the opposite side is called a *median*.

Perimeter

Instruction: The distance around a closed curve is the perimeter. You find the perimeter by adding the lengths of the sides. If the legs of the triangle below measure 20 and the base measures 10, then the perimeter of the triangle is 50 because 20 + 20 + 10 = 50. (To find the **semiperimeter** of a triangle, simply divide the perimeter by 2.)

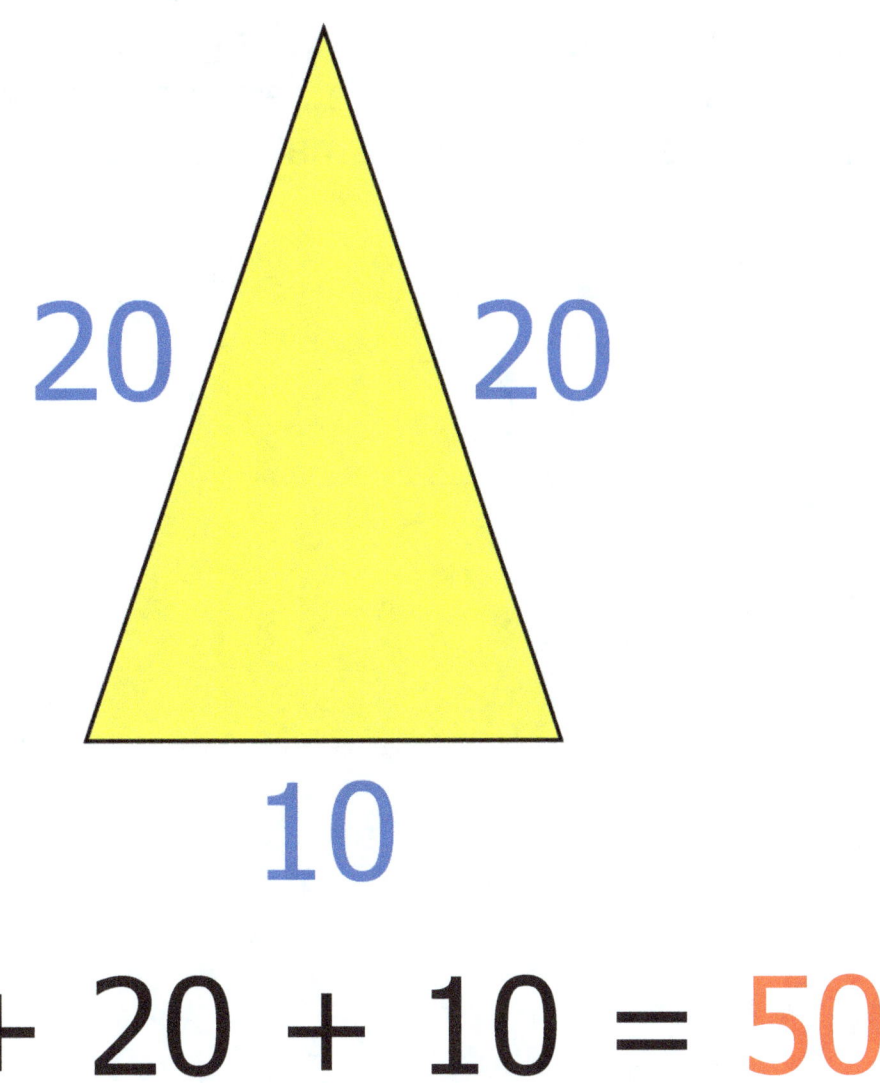

20 + 20 + 10 = 50

If triangle *ABF* ≅ *BCD* ≅ *DEF* ≅ *BDF*, then how many times larger is the perimeter of △*ACE* than that of △*DEF*? Record the factor in the space provided.

Instruction: Since the problem tells us that the smaller triangles are all congruent, then their sides must all be the same lengths. Let's say that all the sides of the little triangles equal 1. This would mean that the smaller triangles have a perimeter of 3 and the larger triangle has a perimeter of 6. Therefore, since 3 × 2 = 6, the perimeter of the larger triangle is twice the perimeter of the smaller triangle.

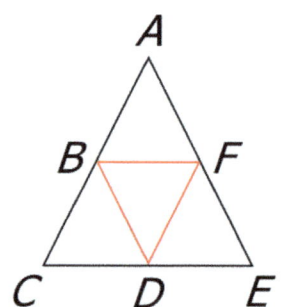 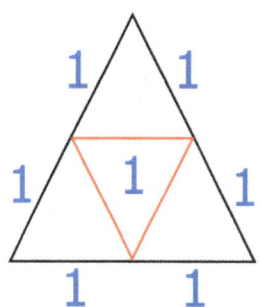

2 times larger

Chapter 4

Circles and Spheres

(Suggested Grades: 8th and 10th)

Teacher instructions: Using *70 Times 7 Math: Electronic Textbook for Teachers (Geometry for Middle and High School Students),* ask students to identify any missing answers for you to write on the screen. Please note that since the answers are provided in student textbooks, they should have them closed during this time. Student textbooks can also be used as a key for the teacher's benefit.

Circles

Instruction: A circle is defined as the set of all points in a plane equidistant (the same distance) from the center point. In the figure to the left below, the red point represents the center point, while the blue points represent the set of points that are the same distance from this center point. Some books use the word *locus* in the definition of a circle, which is a mathematical term used to describe the location of the points.

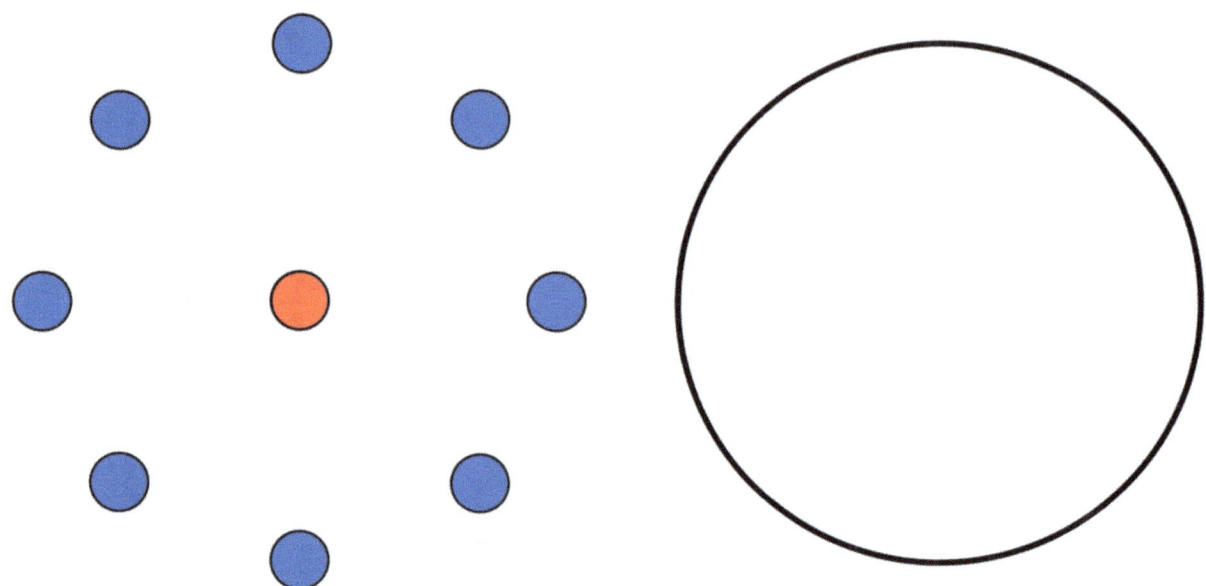

Instruction: Circles are named by the center point. Thus, the circle below is named ⊙A (circle *a*).

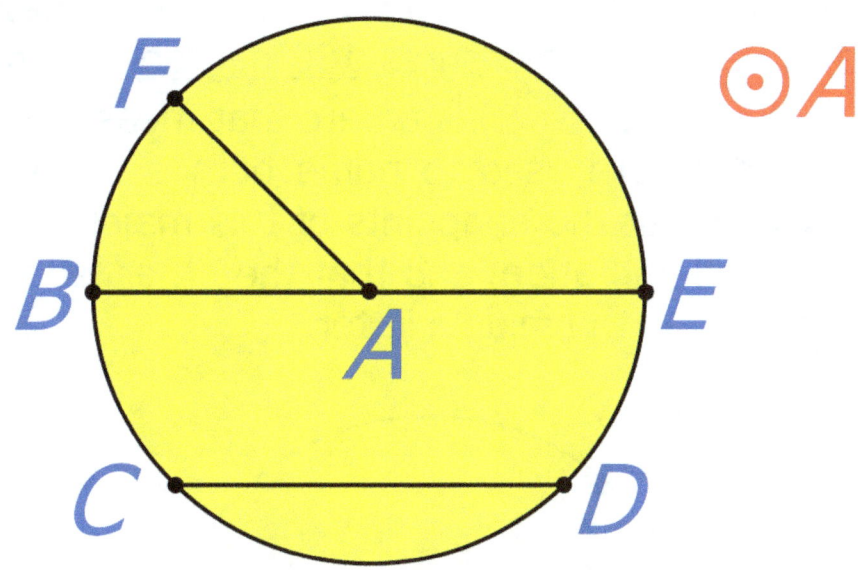

chord: $\overline{CD}$, $\overline{BE}$

A segment having both endpoints on the circle is a <u>chord</u>; $\overline{CD}$ and $\overline{BE}$ are chords of this circle.

diameter: $\overline{BE}$

Besides being a chord, $\overline{BE}$ is also the diameter (*d*) of this circle. A chord that passes through the center of the circle is a <u>diameter</u>.

radius (*r*): $\overline{AB}$, $\overline{AE}$, $\overline{AF}$

A line segment from the center of a circle to a point on the circle is a radius (plural *radii*). The radius is ½ the length of the diameter, and all radii of a circle have the same length. Segments *AB* and *AE* are radii.

Teacher instructions: Use an electronic pen to insert additional chords and radii and let students identify them by name.

Instruction: The measure of a **circle** is 360°. (If you add all the arcs of a circle together, the sum will always be 360°.) One **arc** of this circle is the curve $\overset{\frown}{DE}$; it is a **minor arc** because it measures less than 180°, and it is named with two letters. A **semicircle** is a half-circle; a semicircle measures 180 degrees. A **major arc**, such as $\overset{\frown}{CBD}$ (also called $\overset{\frown}{CED}$) is an arc that measures more than 180°. Three letters are used to name both a major arc and a semicircle. Notice that the endpoints of this major arc (C and D) are both used in the notation, and that the other letter inside the arc (B or E) is used as the middle letter.

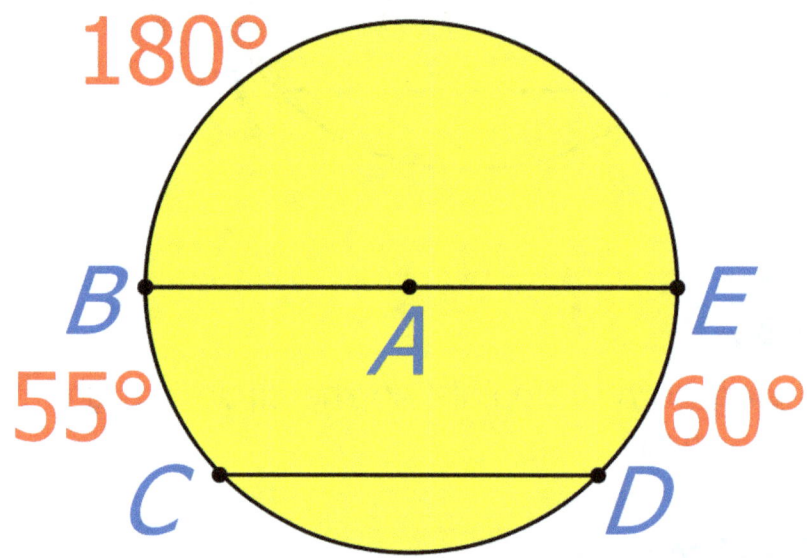

If $\overset{\frown}{BC}$ measures 55° and $\overset{\frown}{DE}$ measures 60°, what does $\overset{\frown}{CD}$ measure?

Solution: The semicircle measures 180°; 180 − 60 − 55 = 65, so $\overset{\frown}{CD}$ measures 65°.

In the circle below, X is the center of the circle, and $\overline{WX} \cong \overline{YZ}$ (they are congruent). Find the degree measure of angle ZXY and of ∠WXZ?

∠ZXY = 60°

∠WXZ = 120°

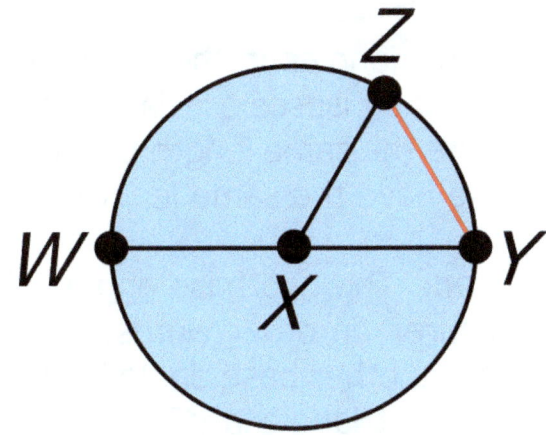

Instruction: We know that all radii of a circle are congruent. Triangle XYZ has to be an equilateral triangle because all of its sides are the same length—the two radii of the triangle ($\overline{XZ}$ and $\overline{XY}$) are the same length, and we are told that side $\overline{YZ}$ is the same length as the radius $\overline{WX}$ (they are congruent). Recall that all the angles of an equilateral triangle measure 60°. Thus, ∠ZXY measures 60°.

If ∠ZXY measures 60 degrees and together ∠WXZ and ∠ZXY must equal 180 degrees since they are supplementary angles (together they form a line), then the measure of ∠WXZ has to measure 120° (180 − 60 = 120).

In the figure below, P is the center of the circle. If UV > VW, then is c > d or is c < d?

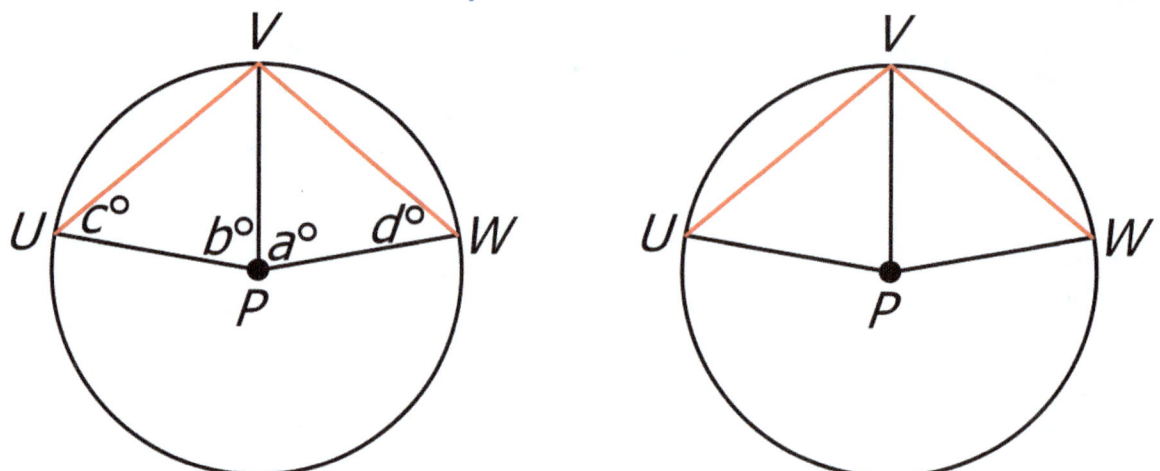

Instruction: See the circle to the left. The two triangles in the circle are isosceles triangles. Recall that at least two sides of an isosceles triangle are the same length. We know that two of its <u>sides</u> are the same length because they are radii of the circle, and all radii of a circle have the same length.

In the illustration, the base of each triangle is red. The two base <u>angles</u> of an isosceles triangle are always the same measure. In other words, since one of the base angles of the first triangle is $c°$, its other base angle is also $c°$. And since one of the base angles of the second triangle is $d°$, its other base angle is also $d°$. Together, the three angles of each triangle will add up to 180°. Thus, we could write

$c + c + b = 180$ (the measure of a triangle), and
$d + d + a = 180$.

Recall that in a triangle, the largest angle will always be on the opposite side of the longest side. Therefore, since the instructions tell us that UV is a longer side than VW, we also know that the vertex angle *b* will be a greater angle than the vertex angle *a*.

Let's assume that the greater vertex angle *b* measures 80° ($b = 80°$). If this is true, then both ***c*'s** = 50° because 50 + 50 + 80 = 180. (Record these values in the image to the right.) Now assume that the lesser vertex angle *a* measures 70° ($a = 70°$). If this is the case, then both ***d*'s** would measure 55 because 55 + 55 + 70 = 180. (Record these values in the image to the right.) Thus, since 50 < 55, then c < d.

The two chords in the circle below are parallel. If the measure of the shorter chord is 12, the measure of the longer chord is 16, and the radius of the circle is 10, then the distance between the chords is what? (Each measurement is in inches.)

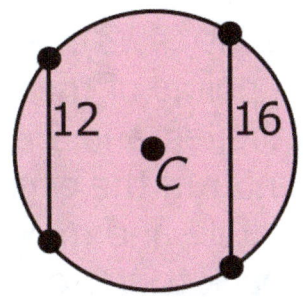
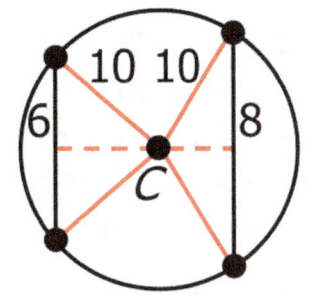

This figure is not drawn to scale.

Solution: Draw in the radii from the center of the circle to the endpoints of the chords, and you form two isosceles triangles (see the figure to the right). If you draw in the height of both isosceles triangles (the perpendicular bisectors), you will form four right triangles. This means that you can use the Pythagorean theorem to find the height of both isosceles triangles. Add these heights together, and you have the distance between the two chords.

From the circle on the right, you can see that one of the legs from the first right triangle is 6 in. because $12 \div 2 = 6$.

$6^2 + \_\_^2 = 10^2$ $36 + \_\_^2 = 100$ $100 - 36 = 64, \sqrt{64} = 8$

The height of one triangle is 8 inches.

One of the legs from a right triangle on the opposite side of the circle's center point is 8 in. because $16 \div 2 = 8$.

$8^2 + \_\_^2 = 10^2$ $64 + \_\_^2 = 100$ $100 - 64 = 36, \sqrt{36} = 6$

The height of the other triangle is 6 inches. The distance between the two chords is 14 in. because $8 + 6 = $ **14**.

The circle in the (x, y) coordinate plane below passes through the origin at (0, 0). If the center point of the circle is at (4, 3), then identify three other known points that the circle passes through. Also, identify the radius of the circle.

Instruction: Since the center of the circle is at (4, 3), we can use these numbers to help us identify other known points of the circle. We start at the center of the circle and move 4 to the left and 3 up; we can then draw in a radius that extends from the center of the circle to point (0, 6). We then go back to the center of the circle and move 4 to the right and 3 up to point (8, 6). Finally, we go back to the center of the circle and move 4 to the right and 3 down to point (8, 0). Note that you can refer to page 253 if you need assistance identifying the points.

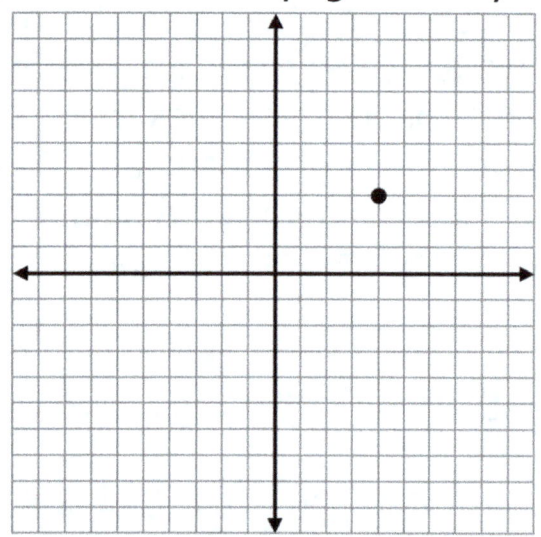
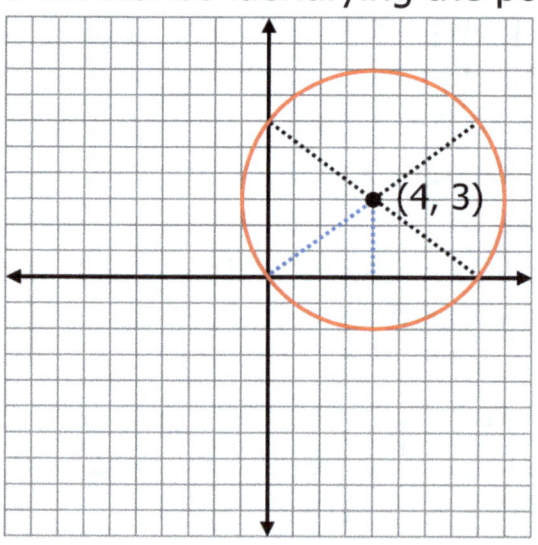

Instruction: To find the radius, we need to draw a right triangle. We make the hypotenuse by drawing a line from the center point to the origin. We then draw a leg of the triangle by drawing a line from the center point to the x-axis at point (4, 0). Now we can see that the longer leg is 4 and the shorter leg is 3, so we use the Pythagorean Theorem to find the length of the hypotenuse (which is also a radius of the circle).

Radius: 5 (**Solution:** $4^2 + 3^2 =$ ___, $16 + 9 = 25$, $\sqrt{25} = 5$)

tangent

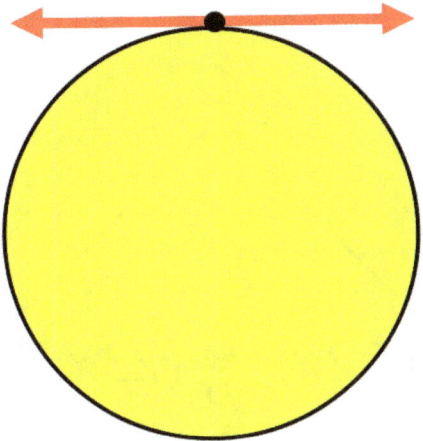

Instruction: A tangent line touches a circle in only <u>one point</u>.

secant

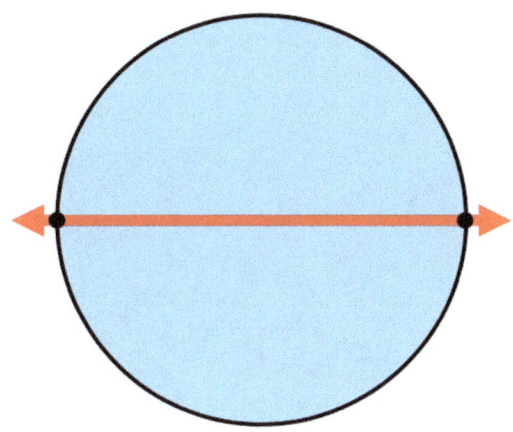

Instruction: A secant line intersects a circle in <u>two points</u>.

inscribed triangle

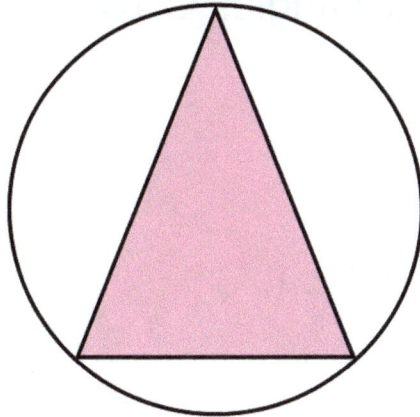

Instruction: The inscribed triangle has no tangent lines (none of the lines touch the circle).

circumscribed triangle

Instruction: The circumscribed triangle shows three points of tangency. The inscribed triangle shows none.

In the figure below, quadrilateral *ABCD* is inscribed in ⊙*P*. Find *x*, $m\angle A$, and $m\angle C$.

x = **4**

$m\angle A$ = **88°**

$m\angle C$ = **92°**

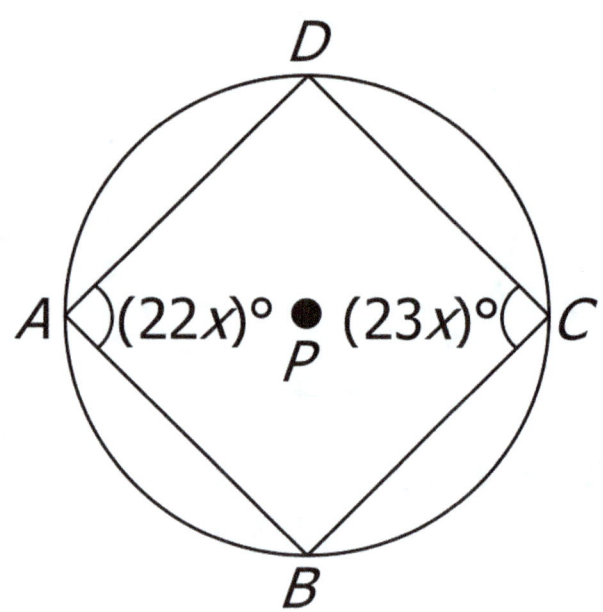

Instruction: We first solve for *x*. If a quadrilateral is inscribed in a circle, then the sum of the opposite angles is 180°.
    22*x* + 23*x* = 180
    45*x* = 180
    x = **4**

The value of *x* is multiplied by 22 to find $\angle A$.
    $m\angle A$ = 22*x*
    22(4) = 88, $m\angle A$ = **88°**
Finally, the value of *x* is multiplied by 23 to find the value of $\angle C$.
    $m\angle C$ = 23*x*
    23(4) = 92, $m\angle C$ = **92°**

Recall that if a quadrilateral is inscribed in a circle, then the sum of the opposite angles is 180°. Compare y and z.

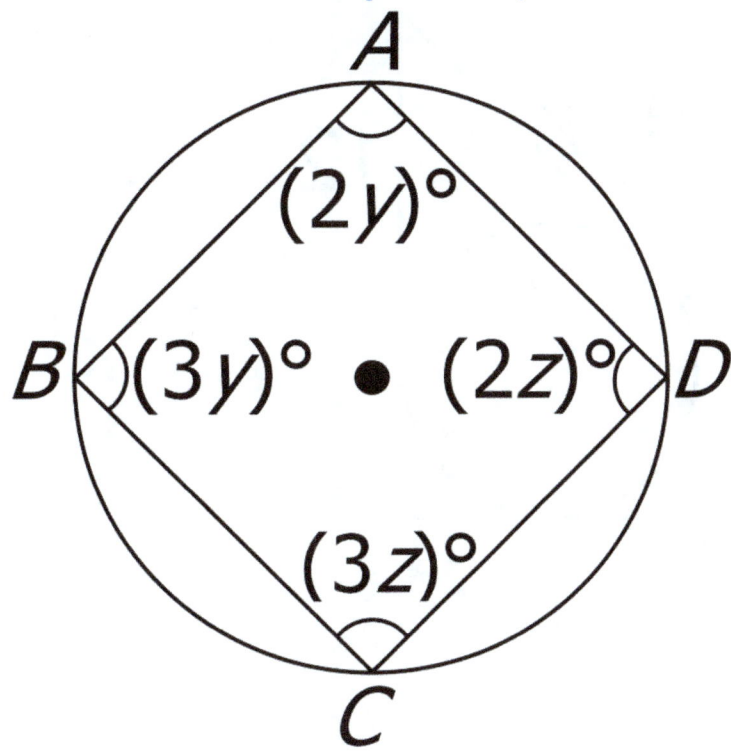

$3y + 2z = 180$ and $2y + 3z = 180$
$3y + 2z = 2y + 3z$
$y = z$

The circumscribed triangle shows three points of tangency. Find x and y.

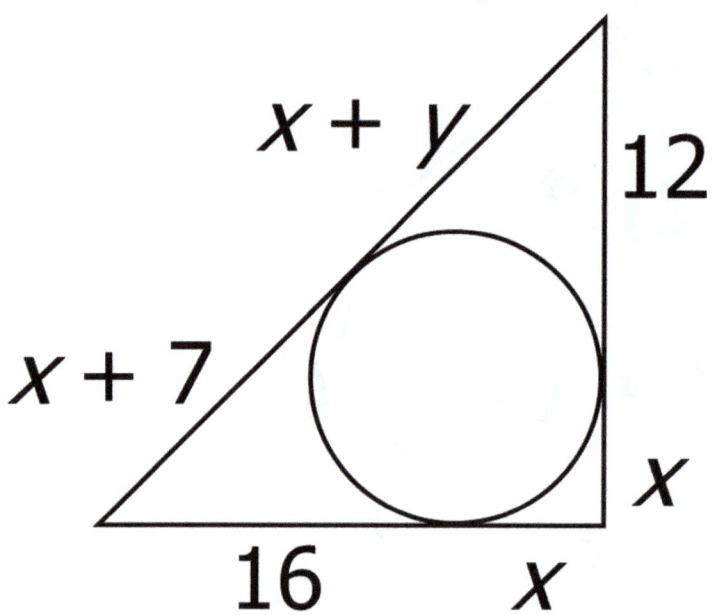

$x = 9$

$y = 3$

Solution:
$x + 7 = 16$, $x = 9$
$x + y = 12$, $9 + y = 12$, $y = 3$

The circumscribed triangle shows three points of tangency. Find x, y, and z.

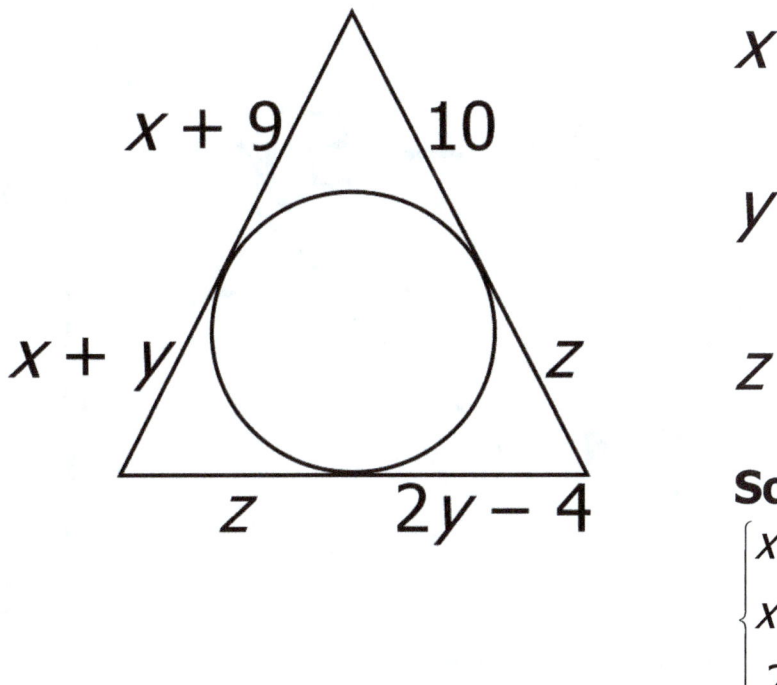

$x = 1$

$y = 5$

$z = 6$

Solution

$$\begin{cases} x + 9 = 10 \\ x + y = z \\ 2y - 4 = z \end{cases}$$

Looking at the equation $x + 9 = 10$, it is obvious that $x = 1$.

$x + 9 = 10$
$x = \mathbf{1}$

Since $x + y$ and $2y - 4$ both equal z, we can record the equation below (notice that x was replaced with 1) and then find the value of y.

$1 + y = 2y - 4$
$\mathbf{5} = y$

(The 4 on the right side of the equation was added to the 1 on the opposite side of the equation to get **5**. The y on the left side of the equation was subtracted from $2y$ on the opposite side of the equation to get **y**.)

Now that we know the value of x and of y, we can find the value of z in the equation $x + y = z$.

$x + y = z$
$1 + 5 = 6$
$z = \mathbf{6}$

The circumscribed triangle shows three points of tangency. Find x and y.

$x = 9$

$y = 3$

Solve:

$$\begin{cases} 4x - 5y = 21 \\ x + 5y = 24 \end{cases}$$

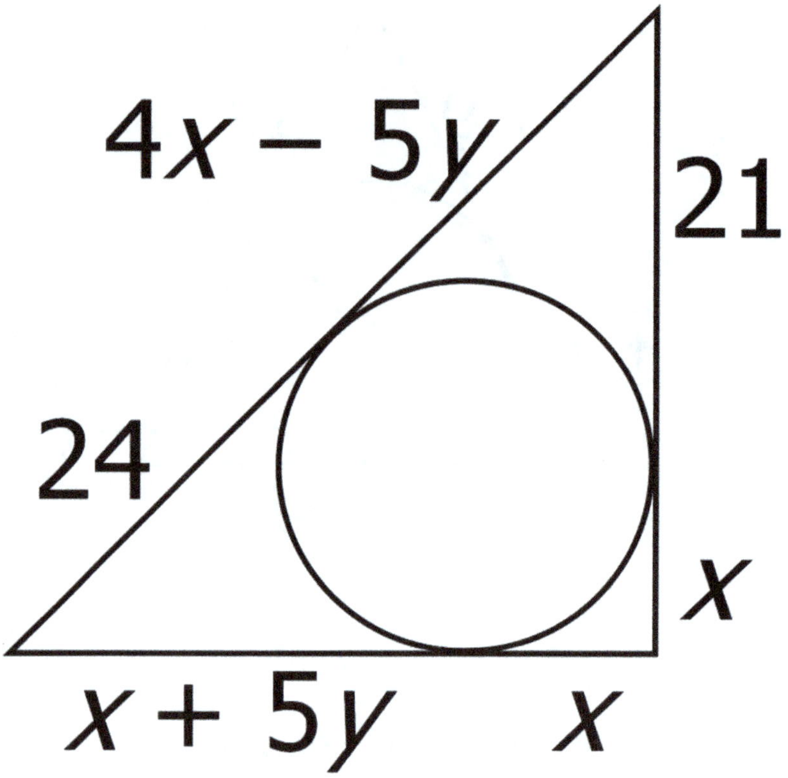

Use elimination to solve.

$$\begin{array}{rl} 4x - 5y &= 21 \\ \underline{x + 5y} &\underline{= 24} \\ 5x \quad\quad &= 45 \\ x &= \mathbf{9} \end{array}$$

$x + 5y = 24$ $\quad$ $9 + 5y = 24$ $(24 - 9 = 15)$ $\quad$ $5y = 15$, $y = \mathbf{3}$

Note: We can check the answers by substituting the value we find for x and for y in each equation.

$4(\mathbf{9}) - 5(\mathbf{3}) = 21$

$\mathbf{9} + 5(\mathbf{3}) = 24$

Instruction: The formula used to find the measurement of an angle formed when two lines (secants or tangents) intersect a circle and intersect each other <u>outside the circle</u> is

(larger arc − smaller arc) ÷ 2.

The measurement of the smaller arc inside the angle is subtracted from the larger arc inside the angle before dividing by 2. If the smaller arc in the first figure below measures 60° and the larger arc measures 112°, x measures 26° because (112 − 60) ÷ 2 = 26.

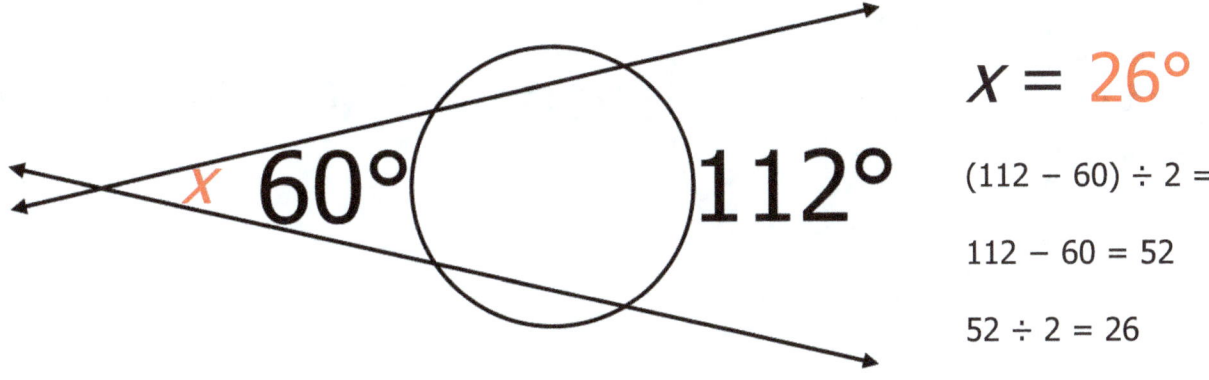

x = 26°

(112 − 60) ÷ 2 =

112 − 60 = 52

52 ÷ 2 = 26

Instruction: The figure above shows four arcs, but the figure below shows only two. Thus, if we know the measurement of one of the arcs but not the other, we can still find it by subtracting the measurement of the known arc from 360 (the measurement of the distance around a circle). Then we can use the formula above to find the measurement of the angle. If the larger arc measures 210°, the other arc measures 150° (360 − 210 = 150), and the angle measures 30° (210 − 150 = 60, 60 ÷ 2 = 30).

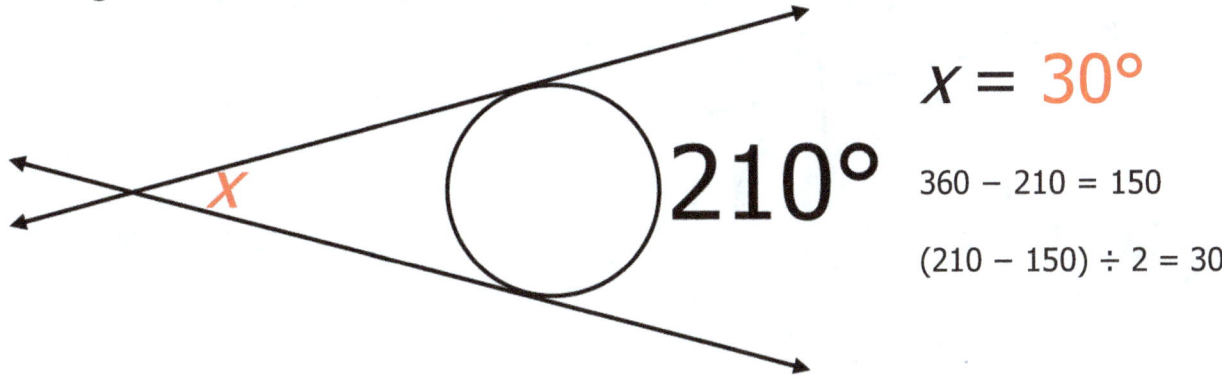

x = 30°

360 − 210 = 150

(210 − 150) ÷ 2 = 30

Instruction: The formula used to find the measurement of the vertical angles formed when two secants intersect each other <u>inside a circle</u> is

$$(\text{arc} + \text{arc}) \div 2.$$

You add the arcs (the two that are inside the vertical angles you are looking for the measurement of) and divide by two.

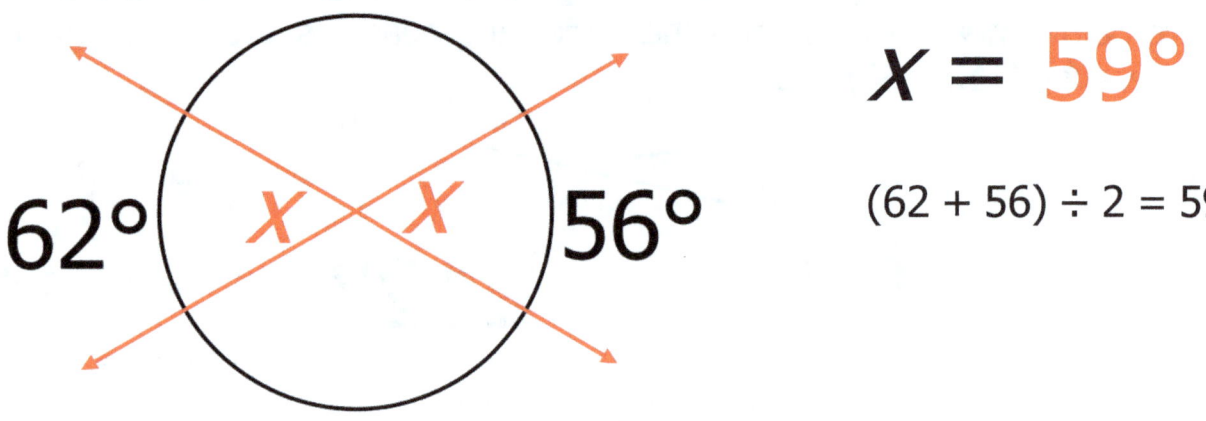

$x = 59°$

$(62 + 56) \div 2 = 59$

Instruction: The formula used to find the measure of an angle formed when two lines intersect a circle and intersect each other <u>on the circle</u> is

$$\text{arc} \div 2.$$

The arc inside the angle you are looking for the measurement of is divided by two.

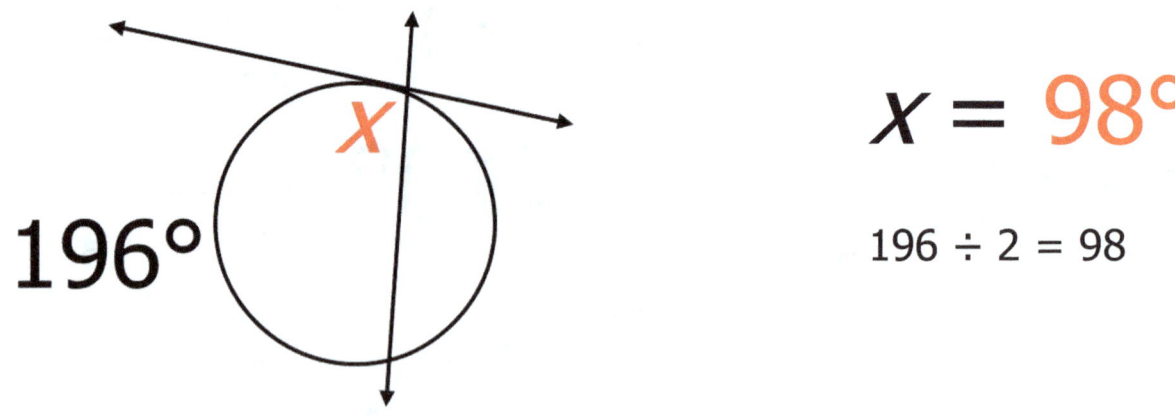

$x = 98°$

$196 \div 2 = 98$

Teacher instructions: The rules from the two previous pages for finding the measurement of an angle formed when two lines intersect a circle are summarized below. Use an electronic pen to insert realistic arc measurements and let students find the measurement of *x*.

1) **Two lines intersect each other <u>outside the circle</u>:**
   (larger arc − smaller arc) ÷ 2

2) **Two lines intersect each other <u>inside the circle</u>:**
   (arc + arc) ÷ 2

3) **Two lines intersect each other <u>on the circle</u>:**
   arc ÷ 2

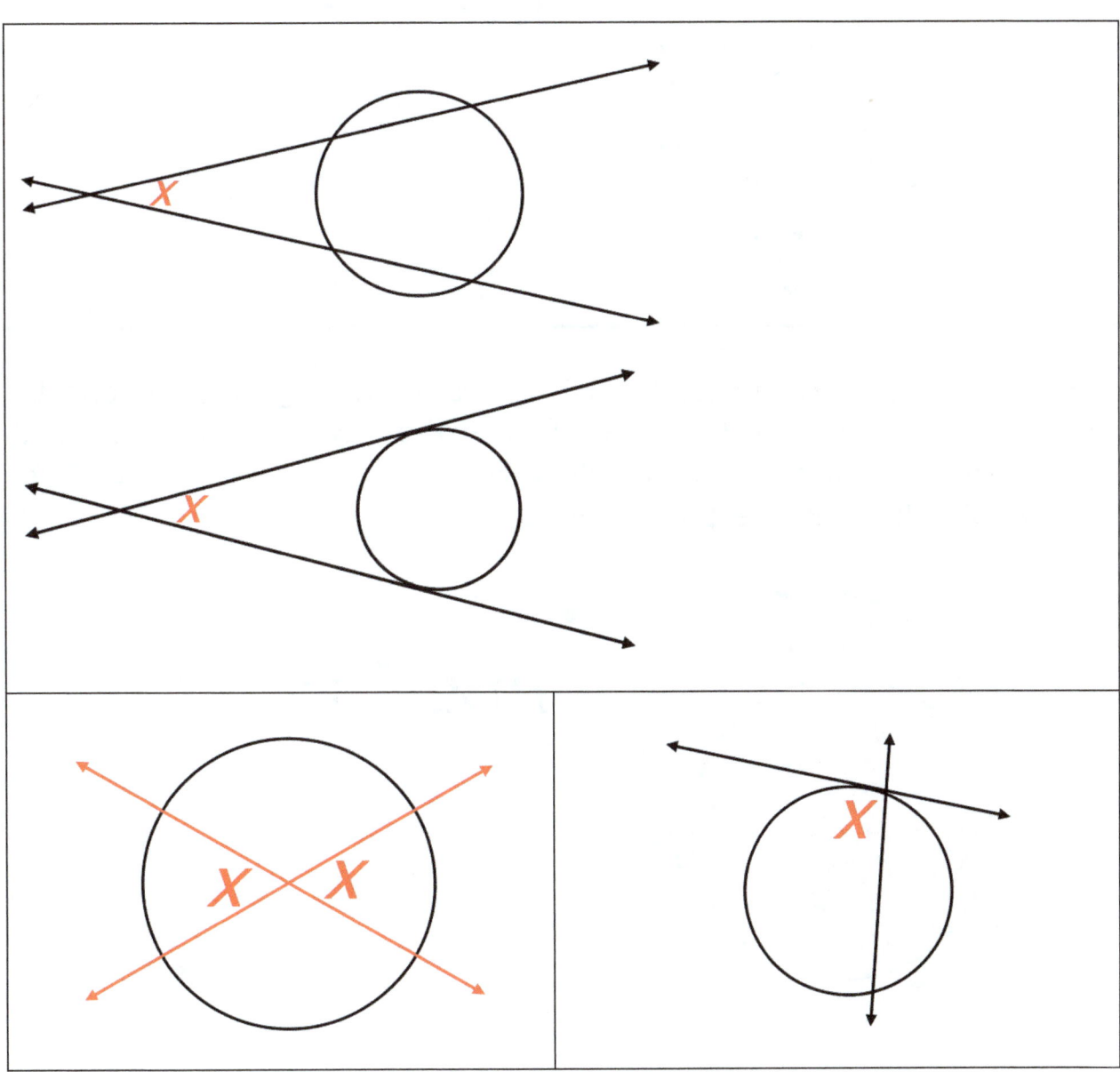

157

Central and Inscribed Angles

Central angle: The vertex of a *central angle* is in the center of the circle, and its two sides intersect the circle. A central angle has the same degree measure as the unique arc it forms. If $m\overset{\frown}{AC} = 105°$, then $m\angle B$ also equals 105°.

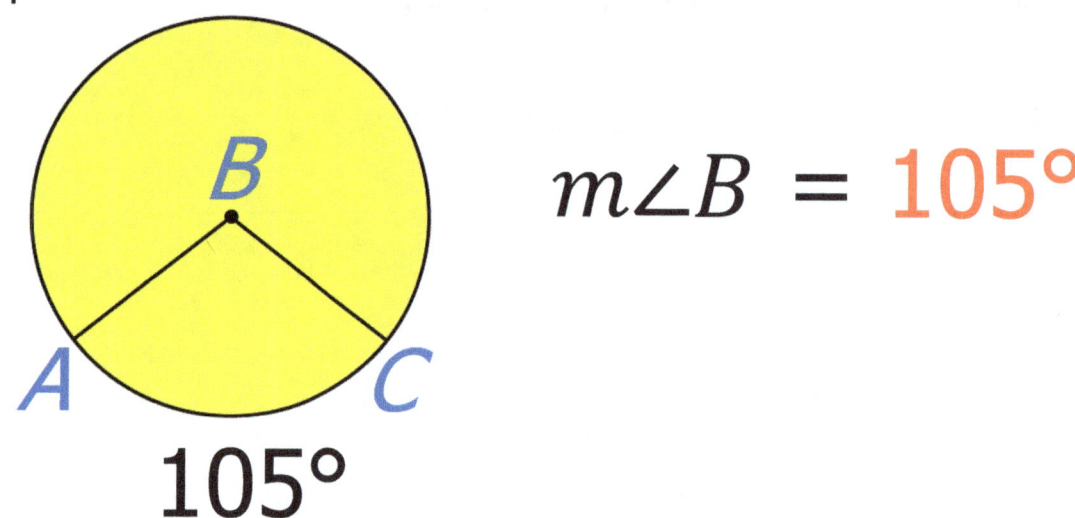

$m\angle B = 105°$

105°

Inscribed angle: The vertex of an *inscribed angle* is on the circle, and its two sides are chords of the circle. To find the measure of an inscribed angle, divide the measure of the unique arc it forms by 2. If $m\overset{\frown}{DF} = 176°$, then $m\angle E = 88°$ because 176 ÷ 2 = 88.

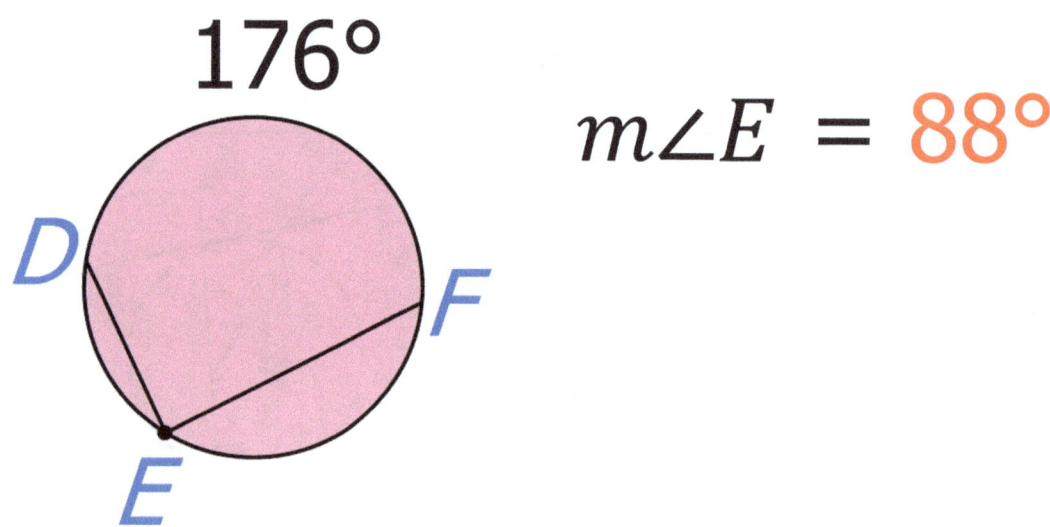

$m\angle E = 88°$

Instruction: If you inscribe an angle inside a semicircle (a half circle), two of its points should be touching the circle on each end of the diameter, and the vertex should also be touching the edge of the circle. Any angle that is inscribed in a semicircle is a right angle and measures 90°.

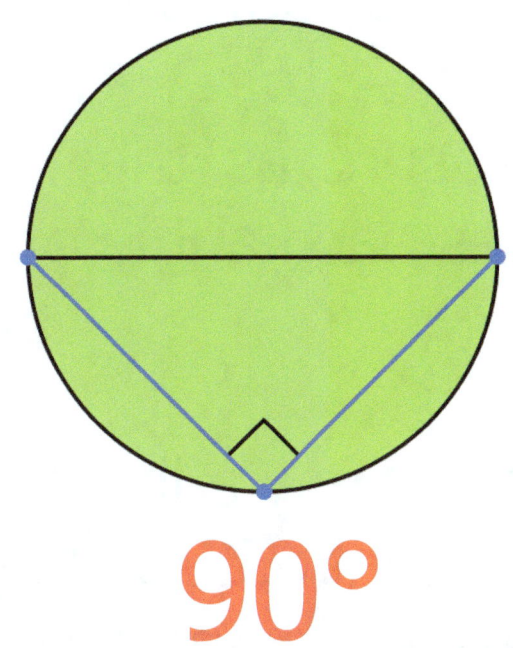

90°

Instruction: Notice the blue tangent line in the figure below. (A tangent line touches a circle in only one point.) If you draw a line (a radius) from the center of the circle to the point of tangency, it will form a right angle and measure 90°. From the illustration below, you can see that the perpendicular distance would be the quickest way to travel from a point to a line, and the quickest way to travel from one point to another point is a straight line.

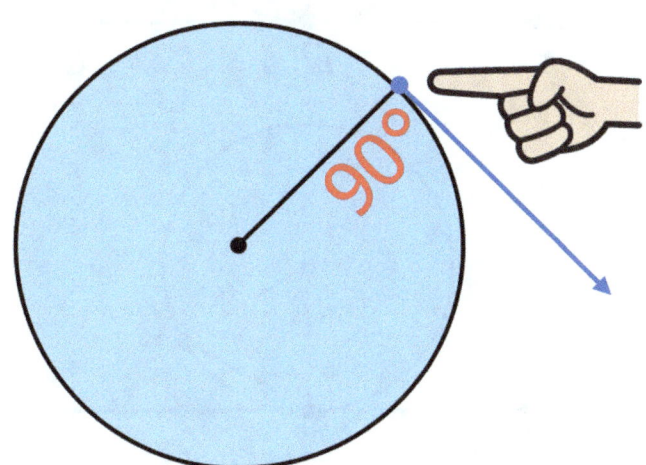

159

Draw a radius from point A to B and from point A to D. If chord BD is the same measure as the radii of the circle, what does each angle of triangle ABD measure?
60°

Now, draw a line from point B to C. If line segment EC is tangent to point D, what does each angle of triangle ACD measure?
(The teacher uses an electronic pen to record the measures in the triangle. Answers are shown in student textbooks.)

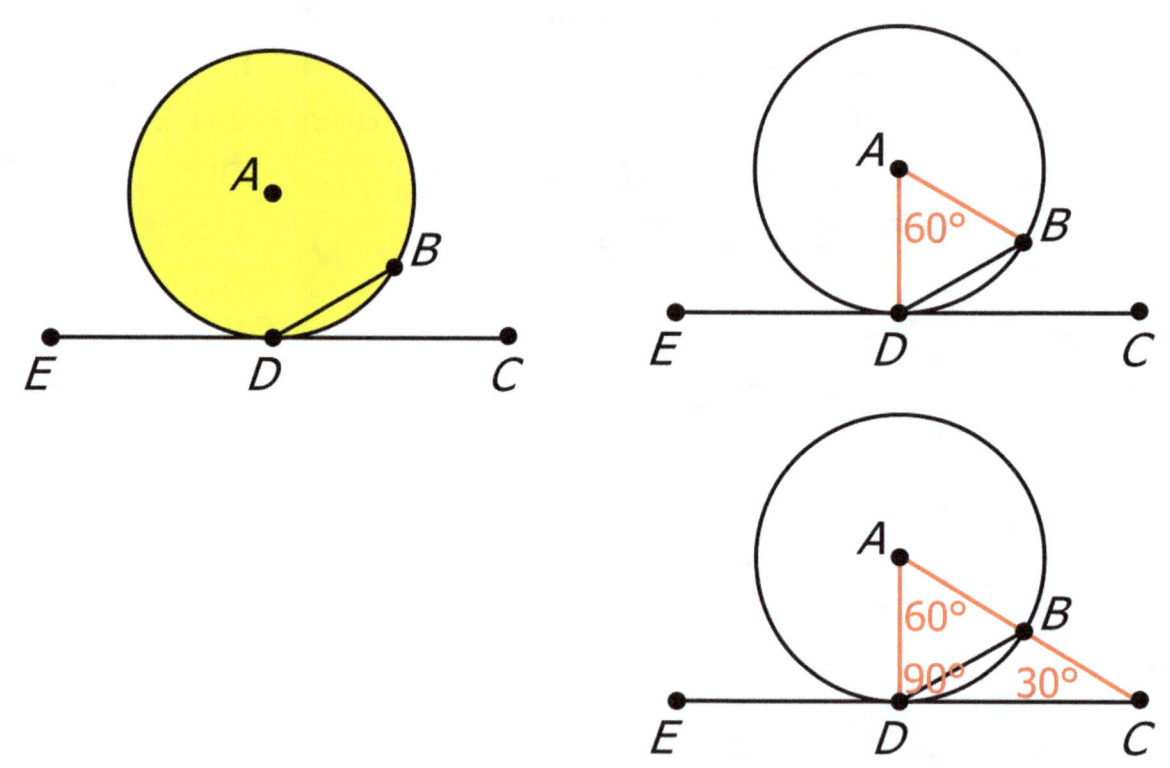

Find *n*.

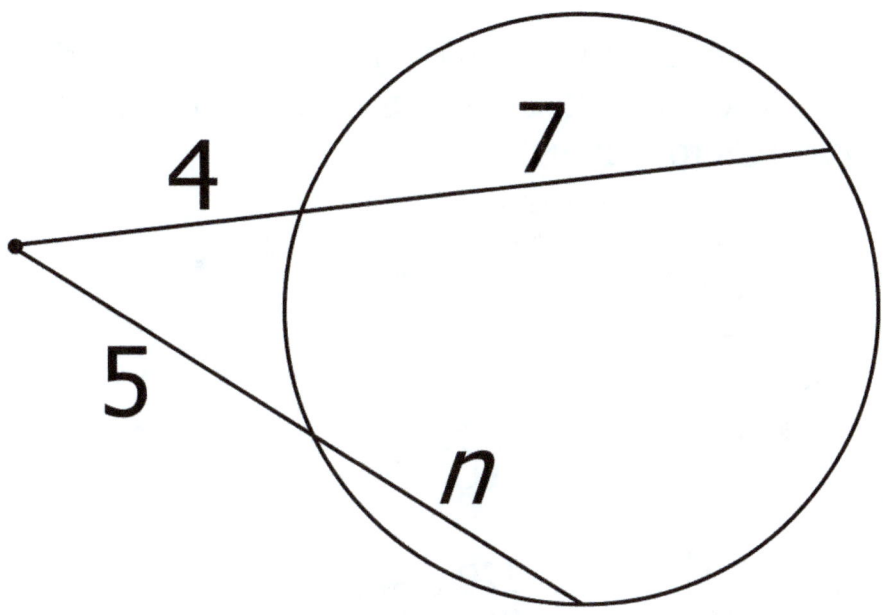

$4(4 + 7) = 5(5 + n)$
$4(11) = 25 + 5n$
$44 = 25 + 5n$
$(44 - 25 = 19)$
$19 = 5n$
$n = \dfrac{19}{5}$ or 3.8

Finding the Length of an Arc

Instruction: The degree of an arc (360° or less) is not the same thing as the length of an arc. The formula to find the length of an arc is recorded below. The *r* stands for radius.

$$\frac{\text{degree of arc}}{360} \times 2r\pi.$$

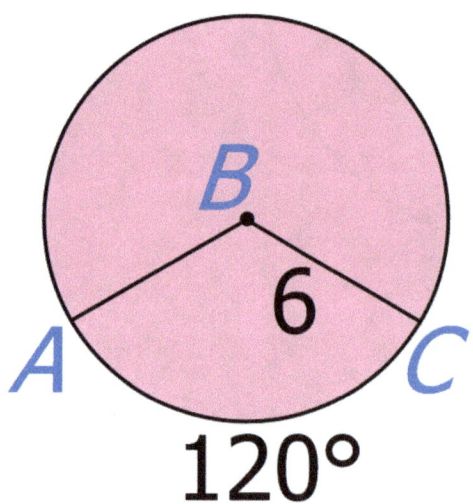

$\widehat{AC} = 4\pi$

$\frac{120}{360} \times 2(6)\pi$

$\frac{1}{3} \times \frac{12\pi}{1} = \frac{12\pi}{3} = \mathbf{4\pi}$

If 120 is the degree measure of $\widehat{AC}$, what is the length of $\widehat{AC}$?

$\frac{120}{360} \times 2(6)\pi$

We reduce 120 over 360 to lowest terms by dividing both numbers by 120. Then we multiply the radius by 2 and record the product 12 as 12 over 1 so that the fractions can be multiplied. Finally, we reduce to lowest terms and record the pi symbol in the answer.

$\frac{1}{3} \times \frac{12\pi}{1} = \frac{12\pi}{3} = 4\pi$ $\widehat{AC} = \mathbf{4\pi}$

Note that a **calculator** can also be used to solve the problem:
120 ÷ 360 × 2 × 6 = 4; insert the pi symbol for 4π.

Notes

What is the length of arc AC?

Hint: Angle *ABC* is a central angle. Recall that a central angle and the arc it forms have the same degree measure. Remember, however, that the degree of an arc and the length of an arc are not the same thing.

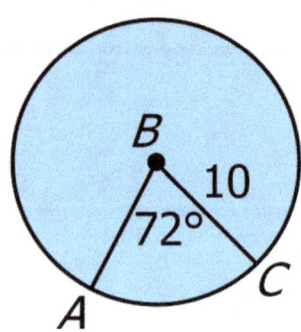

$\overarc{AC}$ = **4π**

Solution:

$\dfrac{72}{360} \times 2(10)\pi \qquad \dfrac{1}{5} \times \dfrac{20\pi}{1} = \dfrac{20\pi}{5}$

20π ÷ 5 = **4π**

What is the length of arc DF if the radius of the circle is 3? Round your answer to the nearest tenth.

Hint: For homework problems like this, you need to know the degree measure of the arc, $\widehat{DF}$ in this example, in order to find the length of the arc. Since $\angle DEF$ is an inscribed angle, the degree measure (not the length) of $\widehat{DF}$ is twice the size of the angle.

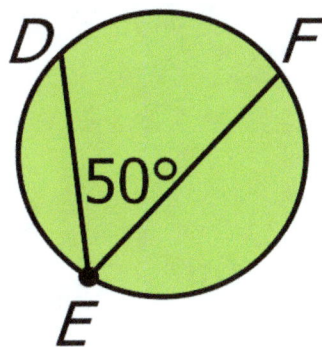

$\widehat{DF} = \dfrac{5\pi}{3}$ or 1.7π

Solution:

$\dfrac{100}{360} \times 2(3)\pi$

$\dfrac{5}{18} \times \dfrac{6\pi}{1} \qquad \dfrac{5}{3} \times \dfrac{1\pi}{1} = \dfrac{5\pi}{3}$

What is the length of arc *AB* if the radius of the circle is 9?

Hint: Since angles *A*, *B*, and *C* form a triangle, we know that ∠*C* must measure 60 because 65 + 55 + 60 = 180. Angle *C* is an inscribed angle, so the degree of $\widehat{AB}$ (not the length) will be twice the size of ∠*C*. Thus, the degree of $\widehat{AB}$ is 120. This information can be used to find the length of $\widehat{AB}$.

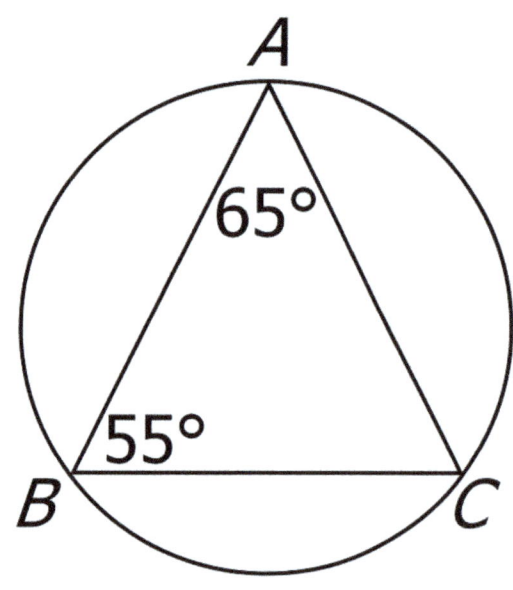

$\widehat{AB} = 6\pi$

$\frac{120}{360} \times 2(9)\pi$

$\frac{1}{3} \times \frac{18\pi}{1} = \frac{18\pi}{3}$ or **6π**

Find the length of arc *BE*. The area of the rectangle *ABCD* is 36 square units.

Solution: Recall that the formula for finding the length of an arc is $\frac{\text{degree of arc}}{360} \times 2r\pi$. Notice that ∠*BAD* is a central angle. Also, recall that a central angle and the arc it forms have the same degree measure. Since the central angle *BAD* is 90° (we know this because rectangles have four right angles), the degree measure of $\overset{\frown}{BE}$ is also 90°, so we can plug this number into the formula. Now we just need to find the length of the radius (*r*). We are told that the area of the rectangle is 36, and we can see that its width is 9. Since the formula for finding the area of a rectangle is "base × height," we just need to find the missing length, which is also its radius (9 × __ = 36, 36 ÷ 9 = 4); the length or radius of the circle is 4.

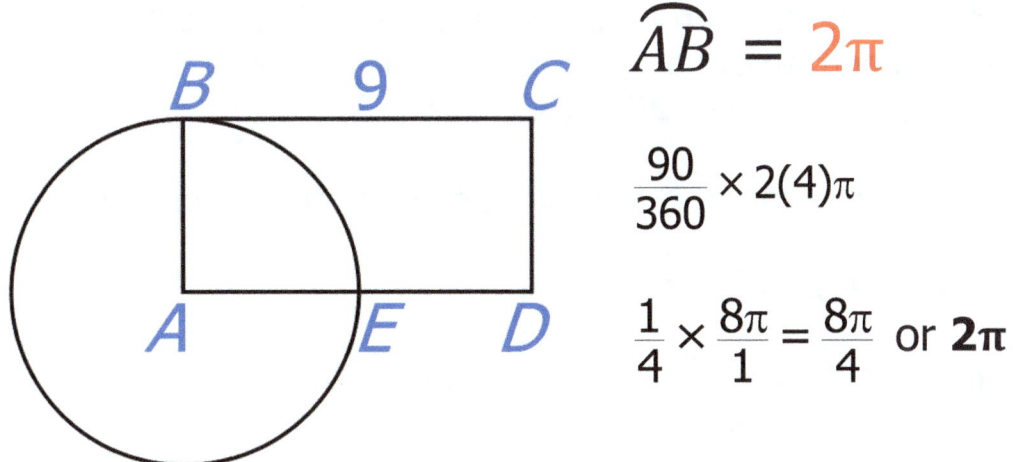

$\overset{\frown}{AB} = 2\pi$

$\frac{90}{360} \times 2(4)\pi$

$\frac{1}{4} \times \frac{8\pi}{1} = \frac{8\pi}{4}$ or **2π**

Note that a **calculator** can also be used to solve the problem: 90 ÷ 360 × 2 × 4 = 2; insert the pi symbol for 2π.

What is the perimeter of the figure below? Note that any angle that appears to be 90° is a 90° angle. Round your answer to the nearest whole number.

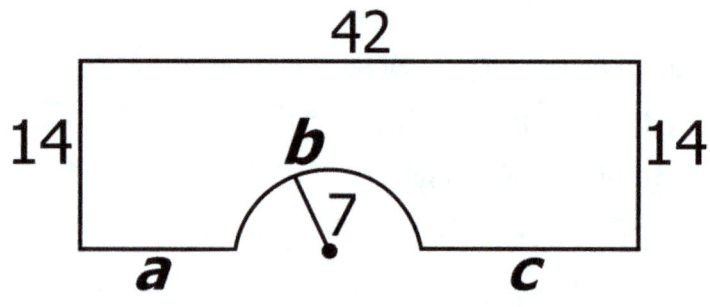

Solution:
$$\frac{\text{degree of arc}}{360} \times 2r\pi$$
$(180 \div 360) \times 2(7)\pi = 7\pi$
$b = \mathbf{7\pi}$

Solution:
$42 - 14 = 28$
$a + c = \mathbf{28}$

Perimeter:
$14 + 42 + 14 + 28 + 7\pi \approx \mathbf{120}$

Instruction: We use the formula for finding the length of an arc to find *b*. (We know that the degree of the arc is 180 because it is a semicircle.)

$\frac{\text{degree of arc}}{360} \times 2r\pi$ $(180 \div 360) \times 2(7)\pi = 7\pi$ $b = \mathbf{7\pi}$

The illustration shows that 42 is the length of the longer side of the rectangle. If we drew a line (or diameter) through the half circle so that there is no gap in the rectangle, its length would be 14 because $7 + 7 = 14$. If we subtract 14 from 42, we find that "$a + c = \mathbf{28}$." ($42 - 14 = 28$)

Now that we have a value for *a*, *b*, and *c*, we can find the perimeter of the rectangle.
  $14 + 42 + 14 + 28 + 7\pi \approx \mathbf{120}$

Using a CASIO fx-9750GIII **calculator**, we plug in
  14 + 42 + 14 + 28 + 7 shift π =
and then round the decimal answer to the nearest whole number.

Note that if the half circle on the illustration was facing out rather than in, it would still be the same perimeter.

pi
π ≈ 3.14

Instruction: The symbol for pi is π. The value of pi is approximately 3.14. The ratio of the circumference of a circle to its diameter will always equal pi.

---

circumference
$d \times π$

Instruction: To find the circumference (which is the distance around a circle or the perimeter of the circle), multiply the diameter by π (pi) on your calculator or by 3.14, which is the approximate value of pi. If you only know the radius, multiply it by 2 before multiplying it by π since it is ½ the length of the diameter.

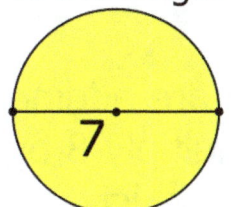

7 × 3.14 = 21.98 or 7π

If the circumference of a circle is 5 cm, what is its diameter? Record the equation you use to find the diameter, and round the diameter to the nearest tenth.

Before writing the equation for problems like this, ask yourself: "What is the formula for finding the circumference of a circle?"
Solution: $d \times 3.14 ≈ 5$ 5 ÷ 3.14 = 1.6 cm when rounded

$$34\pi + 12\pi - 5\pi = 41\pi$$

Instruction: Notice that all the numbers in the equation above are followed by pi. To solve the equation, simply add 34 and 12 together and subtract 5 from the sum to get 41. The pi symbol is then inserted after the 41 to read 41π.

$$8\pi + 2 = 27.13$$
$$8\pi - 2 = 23.13$$

In these problems, only one of the numbers that you are adding or subtracting has the pi symbol. In $8\pi + 2 =$ ___, multiply 8 by 3.14 (the approximate value of pi) and add 2 to the product.

In $8\pi - 2 =$ ___, multiply 8 by 3.14 and subtract 2 from the product. To use your CASIO fx-9750GIII calculator to solve problems like this, see the guide below. You can round the answers to the nearest hundredth.

8 shift π + 2 EXE   and   8 shift π − 2 EXE

$$8\pi \div \pi = 8$$
$8 \div 1 = 8$

$8\pi \div 2\pi = 4$

Instruction: To divide one number with pi by another number with pi, cancel out the pi symbols and divide the numbers. ($8 \div 2 = 4$)

$8\pi \div 2 = 4\pi$

To divide one number with pi by another number without pi, divide the numbers and include the pi symbol in your answer. ($8 \div 2 = 4$, 4π)

$8\pi \times 2 = 16\pi$

If you multiply 8 by pi and then multiply the product by 2, you get a decimal answer because pi ≈ 3.14. If you want to leave your answer *in terms of pi*, simply multiply the two numbers and include pi in the product. ($8 \times 2 = 16$, 16π)

$8\pi \times 2\pi = 157.91$

If you use your calculator to multiply one number with pi by the other number with pi, you find that the answer is 157.91 when rounded to the nearest hundredth.

Area	**Surface Area**
circle	sphere
r² × π	r² × π × 4

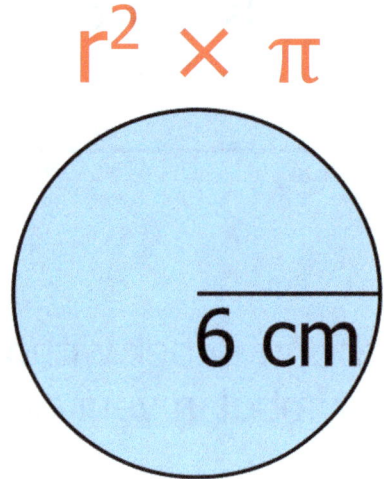 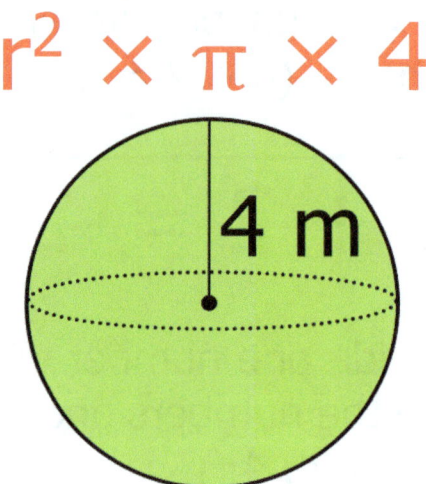

$6^2 = 36$,
$36 \times 3.14 = $ 113.04 cm²
or 36π cm²

$4^2 = 16$
$16 \times 3.14 \times 4 = $ 200.96 m² or
64π m²

Instruction: To find the area of this circle, square the radius (by multiplying it by itself) and then multiply the result by π (pi) or 3.14—the approximate value of pi ($A = \pi r^2$).

If only the diameter was given (the teacher can insert the diameter), we would first need to divide it by two to find the radius to insert in the formula.
($6 \div 2 = 3$, $3 \times 3 = 9$,
$9 \times 3.14 = 28.26$ or 9π cm²)

Instruction: To find the surface area of a sphere, square the radius (by multiplying it by itself); multiply the result by π (pi) or 3.14; then multiply the product by 4 ($A = 4\pi r^2$). If you do not multiply by pi, then keep the pi symbol in your answer (64π instead of 200.96 in the problem above).

You can find the surface area of a sphere by using the formula "$r^2 \times \pi \times 4$." If the surface area of a sphere is 576π cm², what is its radius?

Solution: $r^2 \times \pi \times 4 = 576\pi$, $576 \div 4 = 144$, $\sqrt{144} =$ 12 cm

A large circle has a radius of R cm, and a smaller circle has a radius of $\frac{2}{3}r$ cm. Find the ratio of the <u>area</u> of the smaller circle to that of the larger one.

$$\frac{\text{Small circle}}{\text{Large circle}} = \frac{\pi\left(\frac{2}{3}R\right)^2}{\pi R^2} = \frac{\frac{4}{9}R^2}{R^2} = \frac{4}{9}$$

Finding the Area of a Circle

If the area of the square inscribed in the circle is 16 cm², what is the area of the circle?

Solution: We need to apply what we know about 45°-45° right triangles to find the area of the circle. Recall that one diagonal divides a square into two 45°-45° right triangles. If the area of the square is 16 cm², then each side of the square is 4 cm because you find the area of a square by squaring one of its sides; and, in contrast, you would find the length of one of its sides by finding the square root of the area—$\sqrt{16} = 4$.

Thus, we have the length of the legs (4 cm) of the 45°-45° right triangles that the diagonal divided the square into. And, if we know that the length of the legs is 4 cm, then we also know that the length of the hypotenuse of the triangle is $4\sqrt{2}$ cm. This diagonal can be thought of as the diameter of the circle, so if $4\sqrt{2}$ cm is the diagonal, the radius will be one half its length—$2\sqrt{2}$ cm. This is important because we have to know a circle's radius in order to find its area (the formula for finding the area of a circle is $r^2 \times \pi$.) Therefore, the area of the circle is $2\sqrt{2} \times 2\sqrt{2} \times \pi = 8\pi$ cm².

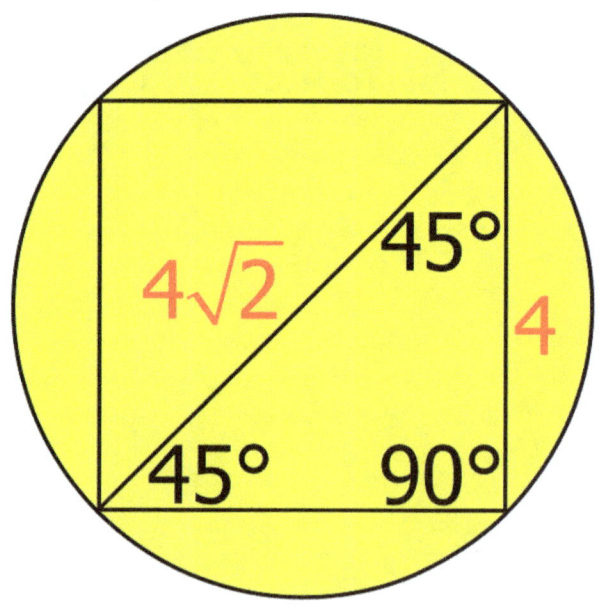

side of square: $\sqrt{16} = 4$ cm

hypotenuse = $4\sqrt{2}$ cm

radius = $2\sqrt{2}$ cm
$$\frac{4\sqrt{2}}{2} = 2\sqrt{2}$$

area of circle: 8π cm²
$2\sqrt{2} \times 2\sqrt{2} \times \pi = 8\pi$

Find the probability of a dart hitting the shaded area of the dartboard below. The radius of the larger circle is 6, and the radius of the smaller circle is 4.

P = 0.56

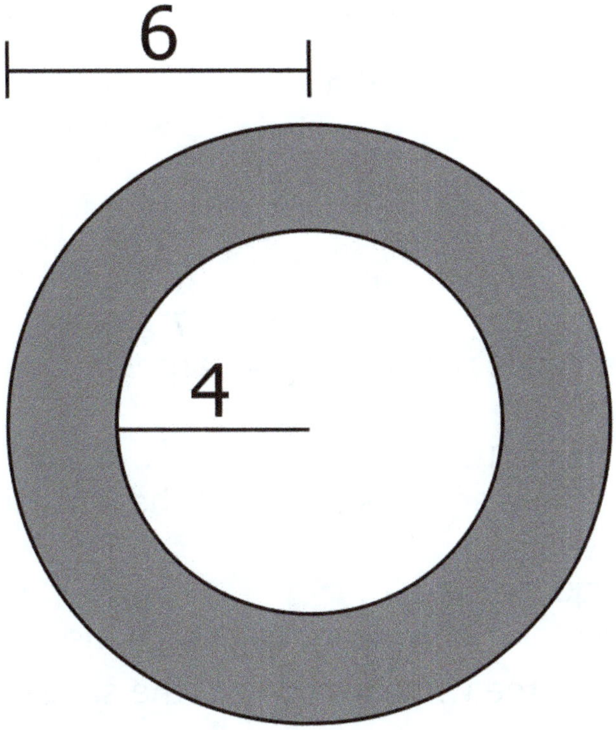

Note: When you are dividing two numbers that both have the pi sign, cancel out the pi signs before dividing.

Solution:

$$P = \frac{(\pi 6^2) - (\pi 4^2)}{\pi 6^2} = \frac{36\pi - 16\pi}{36\pi} = \frac{20\pi}{36\pi} = \frac{5\pi}{9\pi} = \frac{5}{9} = 0.56$$

If only one dart is tossed at the dartboard illustrated, what is the probability that it will land on a white section of the smaller circle?

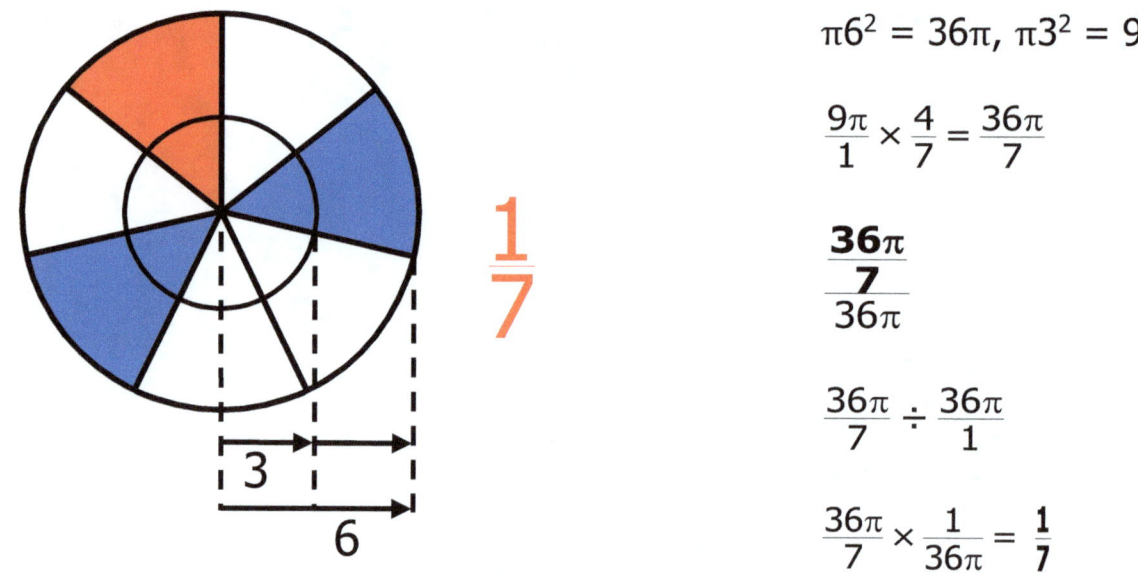

$\pi 6^2 = 36\pi, \pi 3^2 = 9\pi$

$\dfrac{9\pi}{1} \times \dfrac{4}{7} = \dfrac{36\pi}{7}$

$\dfrac{\frac{36\pi}{7}}{36\pi}$

$\dfrac{36\pi}{7} \div \dfrac{36\pi}{1}$

$\dfrac{36\pi}{7} \times \dfrac{1}{36\pi} = \dfrac{1}{7}$

Instruction: First, find the area of the whole dartboard (the larger circle). The diagram shows that it has a radius of 6.

Area of the whole dartboard: $\pi 6^2 = 36\pi$ square units

Now find the area of the smaller circle. The diagram shows that the smaller circle has a radius of 3.

Area of the smaller circle: $\pi 3^2 = 9\pi$ square units

Since the area of the smaller circle is 9π and only 4 out of 7 sections are white, multiply 9π by $\dfrac{4}{7}$ to find the area of the white sections of the smaller circle: $\dfrac{9\pi}{1} \times \dfrac{4}{7} = \dfrac{36\pi}{7}$.

Next, record $\dfrac{36\pi}{7}$ over 36π because $\dfrac{36\pi}{7}$ (the over portion) is the area of the white region of the smaller circle and 36π (the under portion) is the area of the whole dartboard. Then you can divide $\dfrac{36\pi}{7}$ by 36π.

$\dfrac{\frac{36\pi}{7}}{36\pi} \qquad \dfrac{36\pi}{7} \div \dfrac{36\pi}{1} \qquad \dfrac{36\pi}{7} \times \dfrac{1}{36\pi} = \dfrac{1}{7}$

(The 36π's will cancel each other out.)

Area of a Sector (Pie Slice)

$$\frac{\text{central angle}}{360} \times \text{area of circle}$$

Instruction: A sector is shaped like a piece of pie and is bounded by a central angle. The area of a sector is the measure of the central angle divided by 360 and multiplied by the area of the circle. The area of the entire circle below is 36π square units ($6^2 \times \pi = 36\pi$). The area of the sector bounded by the central angle is 9π square units ($90 \div 360 \times 36\pi = 9\pi$).

Area of the circle: 36 square units
   $6^2 \times \pi = 36\pi$

Area of the sector: 9π square units
   $90 \div 360 \times 36\pi = 9\pi$

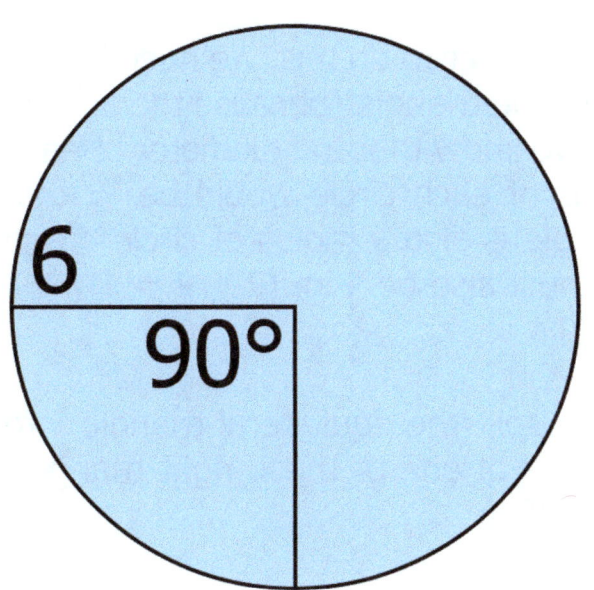

The center point of the three adjacent circles below are X, Y, and Z. If points X, Y, and Z form an equilateral triangle and the diameter of each circle is 4 in, then the area of the shaded region is what? Record your answer as the area of the triangle minus the area of the 3 sectors inside the triangle.

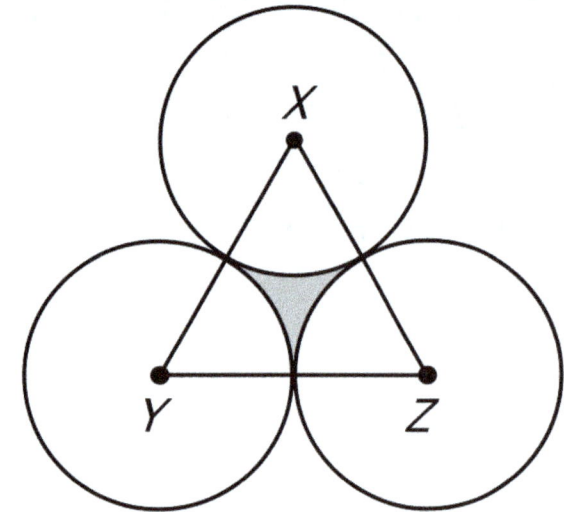

Base of the equilateral triangle: 4 in
First, find the area of the entire triangle. In order to do this, we need to find the length of the base and of the height of the triangle (because the area of a triangle is "base × height ÷ 2"). Since we know that the diameter of each circle is 4 in, we also know that a radius of each circle would be ½ that length. Thus, since the base of the triangle is also a radius of circle Y and of circle Z, then the base of the triangle must also be 4 in. (2 + 2 = 4).

Height of the equilateral triangle: $2\sqrt{3}$ in
To find the height of the triangle, we can split the equilateral triangle into two 30°-60° right triangles. Since the base of one of these right triangles would be 2 in, then the height would be $2\sqrt{3}$ in.

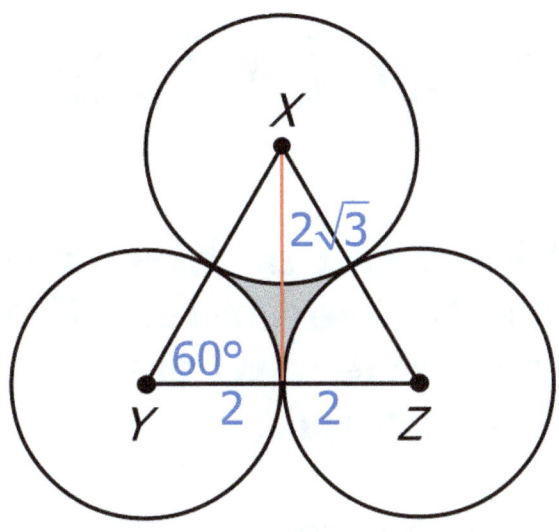

Area of the equilateral triangle: 4√3 in²
Thus, the area of the triangle must be 4√3 in² since 4 × 2√3 ÷ 2 = 4√3.
$$\frac{4 \times 2\sqrt{3}}{2} = \frac{8\sqrt{3}}{2} = 4\sqrt{3}$$

Area of one circle: 4π in²
The area of a circle is "radius² × π." (**Solution:** 2² × π = 4π)

Area of the 3 sectors inside the triangle: 2π in²
Recall that the formula to find the sector (or pie slice) of a circle is
$\frac{\text{central angle}}{360}$ × area of circle.
Use the formula to find the area of one of the sectors; then you can multiply that area by 3 (since there are 3 sectors) and subtract the total from the area of the whole triangle to find the area of the shaded region. Since each angle of an equilateral triangle measures 60°, 60 will be the central angle we plug into the formula. We can reduce 60/360 to 1/6 before multiplying it by the area of the circle (the sector is 1/6 the area of the entire circle).
$$\frac{60}{360} = \frac{1}{6} \qquad \frac{1}{6} \times \frac{4\pi}{1} = \frac{4\pi}{6} = \frac{2\pi}{3}$$
Now multiply the area of one sector by 3:
$$\frac{2\pi}{3} \times \frac{3}{1} \qquad \frac{2\pi}{1} \times \frac{1}{1} = \frac{2\pi}{1} = 2\pi.$$

Area of the shaded region: 4√3 − 2π in²

More about Tangent Lines

Theorem: Exactly two tangent lines can be extended from any point outside a circle to the circle. The distance from the point outside the circle to where each segment touches the circle is always equal.

The theorem above can be used to help answer the following question.

How do we know that line segment *AB* and line segment *AC* are congruent in the figure below?

Because they meet at the same point outside the circle (at point *A*), and they are tangent to the circle at points *B* and *C*.

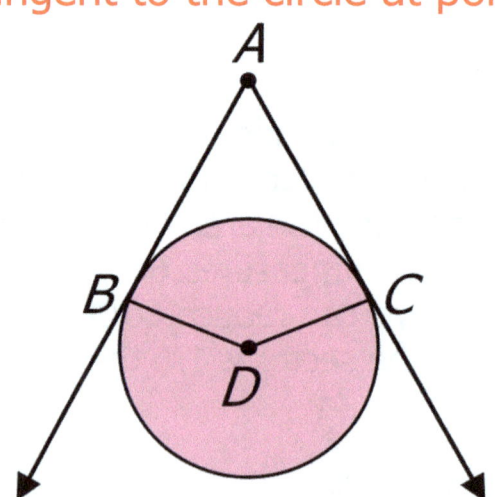

Ray *AB* and *AC* are both tangent to circle *D*. What is the area of quadrilateral *ABDC* if ∠*BAC* is 60° and the radius of the circle measures 4 cm? **$16\sqrt{3}$ cm²**

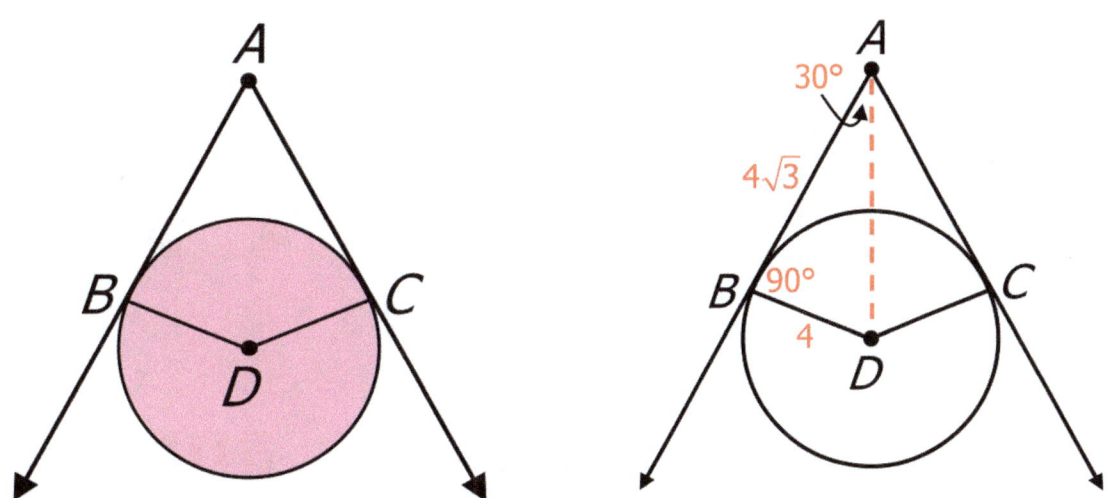

Instruction: Recall that when a tangent line and a radius intersect (as they do at points *B* and *C* above), they form a right angle. If you draw a line from *A* to *D*, then you form two 30°-60°-90° triangles because angle *A* (which was 60°) would split in half leaving you with two 30° angles. Each triangle would also have the 90° angle, so the last angle would have to measure 60° because $90 + 30 + 60 = 180$. Since we need to find the area of quadrilateral *ABDC*, we can simply find the area of one of the congruent right triangles and then multiply it by 2. We are told that the radius (which is also a leg of the 30°-60°-90° triangle) is 4 cm, and thus the other leg has to be $4\sqrt{3}$ cm (see the information about 30°-60° right triangles if you have forgotten the rule). When you multiply the base by the height and divide by 2, we find that the area of one triangle is $8\sqrt{3}$ cm² as seen below.

$$\frac{4 \times 4\sqrt{3}}{2} = \frac{16\sqrt{3}}{2} = \mathbf{8\sqrt{3} \text{ cm}^2}$$

(In the equation above, we multiply 4 by 4 to get 8. Divide 16 by 2 and you are left with $8\sqrt{3}$.)

The area of the quadrilateral (the two triangles together) is $16\sqrt{3}$ cm².
    $8\sqrt{3} \times 2 = \mathbf{16\sqrt{3} \text{ cm}^2}$

Chapter 5

Area and Surface Area

(Suggested Grades: 8ᵗʰ and 10ᵗʰ)

Note to teachers: When studying chapter 5, please also take time to review pages 103, 105, 106, 169, and 172 with your class, as they contain information students need to know when completing certain classwork/homework and test problems for this chapter.

Teacher instructions: Using *70 Times 7 Math: Electronic Textbook for Teachers (Geometry for Middle and High School Students),* ask students to identify any missing answers for you to write on the screen. Please note that since the answers are provided in student textbooks, they should have them closed during this time. Student textbooks can also be used as a key for the teacher's benefit.

Area, Surface Area, Volume

Instruction: Area is the number of square units required to fully cover a flat surface. (If you count the square units in the square below, you will find that the area is 20 square units.) **Surface area** is the number of square units required to cover the outside of a 3-D figure. **Volume** is the number of cubic units required to fill the inside of a 3-D figure. A cubic unit is shaped like a cube, and each of its sides measure one unit. When recording the area or the surface area of a figure, a superscript *2* follows the measurement. When recording the volume of a figure, a superscript *3* follows the measurement.

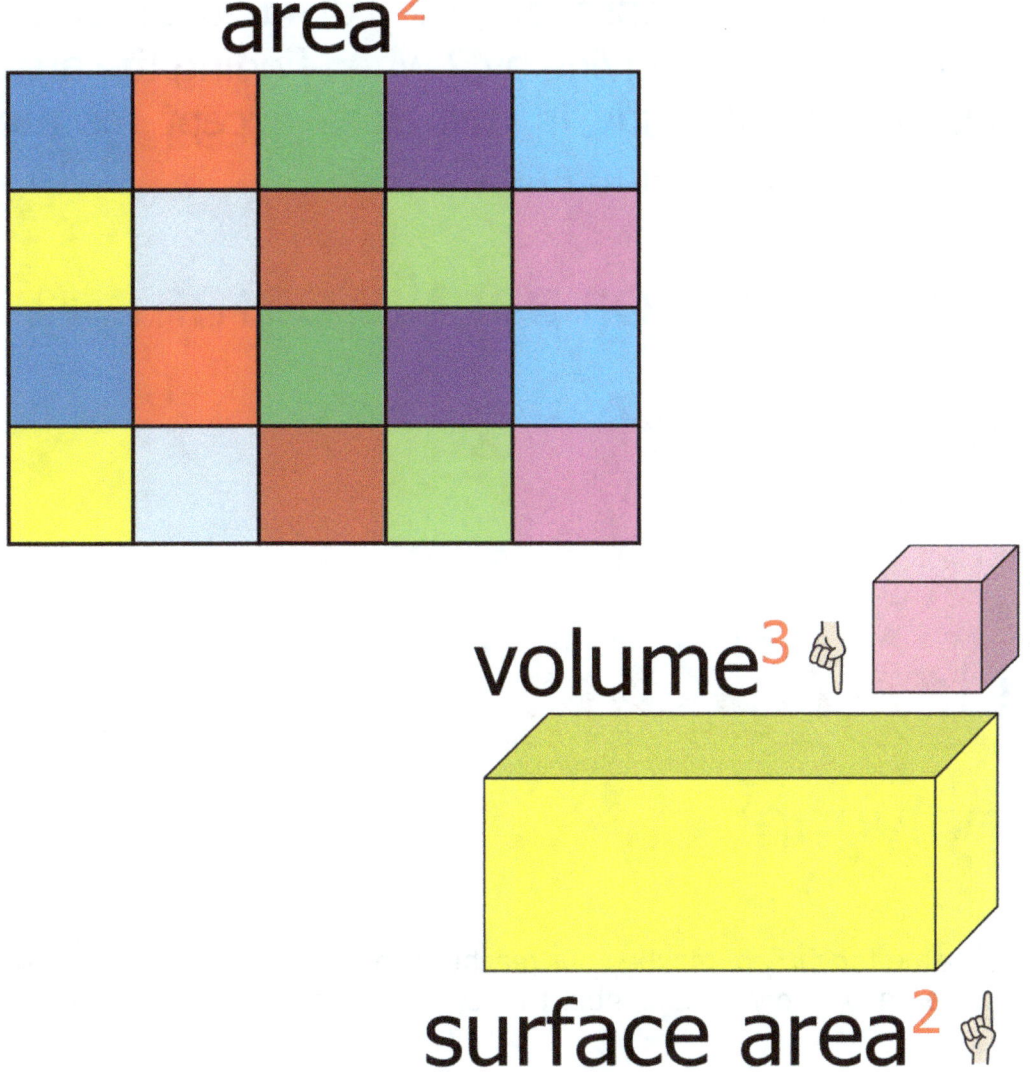

Area Formulas

Instruction: You find the area of a <u>square, a rectangle, and a parallelogram</u> by multiplying the base by the height. To find the area of a <u>rhombus</u>, multiply the base by the height (or multiply the two diameters and divide the product by 2). To find the area of a <u>triangle</u>, multiply the base by the height; then divide the product by 2. To find the area of a <u>trapezoid</u>, you add base 1 and base 2; multiply the sum by the height; then divide the product by 2. To find the area of a regular polygon, multiply the perimeter by the apothem; then divide the product by 2. (**Note:** You do have to divide by 2 when finding the area of all the shapes we will study in this class except for the square, rectangle, parallelogram, and circle.)

square, rectangle, or parallelogram: bh

rhombus: bh or $d_1 d_2 \div 2$

triangle: $bh \div 2$

trapezoid: $h(b_1 + b_2) \div 2$

regular polygon: $pa \div 2$

Hint: *Pa* is another name for *father*. Maybe your teacher can show you a clip of Pa on *Little House on the Prairie* to help you remember the formula for finding the area of a regular polygon!

Area of a square, rectangle, or parallelogram
$A = bh$

Instruction: You find the area of a square, a rectangle, and a parallelogram by multiplying the base by the height. Thus, the area of the square below is 25 cm² (because 5 × 5 = 25 cm²), and the area of the rectangle and parallelogram are 50 cm² (because 10 × 5 = 50 cm²).

SQUARE

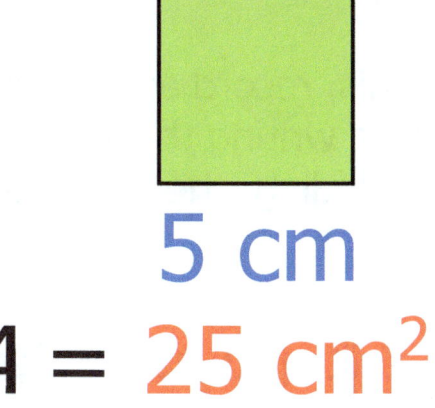

5 cm

$A = 25$ cm^2

Instruction: Although we are only told the length of one of the sides of the square, we know that all the sides of a square are equal.

RECTANGLE

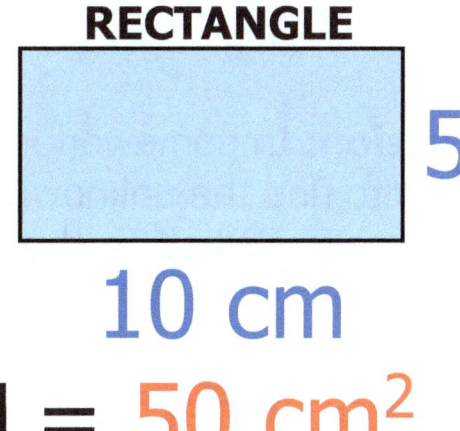

5

10 cm

$A = 50$ cm^2

PARALLELOGRAM

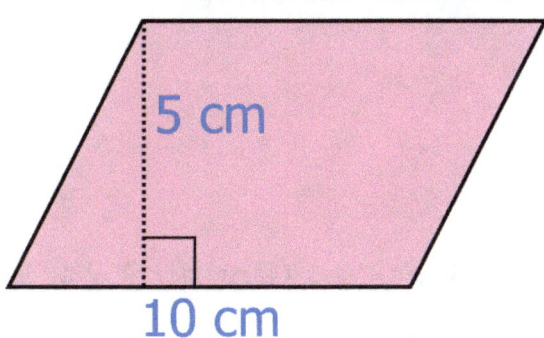

5 cm

10 cm

$A = 50$ cm^2

Instruction: You can illustrate the height of a parallelogram by drawing a perpendicular line (called the *altitude*) from the highest point of the parallelogram to the opposite side.

Area Word Problems

Find the unknown height of a rectangle with an area of 31.5 mm² and a base 4.5 mm.

Record the equation: 4.5 × ___ = 31.5 mm²
(**Solution:** 31.5 ÷ 4.5 = 7)

Height: 7 mm

Instruction: In homework problems like this, record and solve an equation to find the unknown height. Before writing the equation, ask yourself: "What is the formula for finding the area of a rectangle?"

Shana and her husband want to put one coat of stain on the rectangular, wooden floor of their living room. If the floor measures 18 feet by 27 feet and a quart of stain covers 150 square feet, how many quarts should they buy?

4 quarts

Solution: 18 × 27 = 486 ft², 486 ÷ 150 = 3.24 (Round 3.24 up to 4 quarts so that they have enough paint.)

Mom is having new carpet installed in the den. How much carpet is needed for the 24 ft. by 20 ft. room? Convert your answer to square yards since carpet is usually sold by the square yard.

54 yd^2

$24 \times 20 = 480, 480 \div 9 = 53.3$

Instruction: Multiply 24 by 20 to get 480. There are 9 ft² per square yard, so you convert 480 to square yards by dividing by 9. Round 53.3 up to 54 so that she won't be short of carpet.

Use the measurements below to find the outside wall area of a barn. The height of the outer wall is 10 feet.

```
         55 ft.
      ┌─────────┐
42 ft.│         │42 ft.
      └─────────┘
         55 ft.
```

$1,940 \text{ ft}^2$

$55 + 55 + 42 + 42 = 194; 194 \times 10 = 1,940 \text{ ft}^2$

Instruction: Find the perimeter; then multiply the result by 10 since that is the height of the outer wall.

Area of a Rhombus

$$A = bh \text{ or } A = d_1 d_2 \div 2$$

Instruction: To find the area of a rhombus, multiply the base by the height. Another way to find the area of a rhombus is to multiply the length of the two diagonals. Then, divide the product by 2. (Find the area of the rhombus below.)

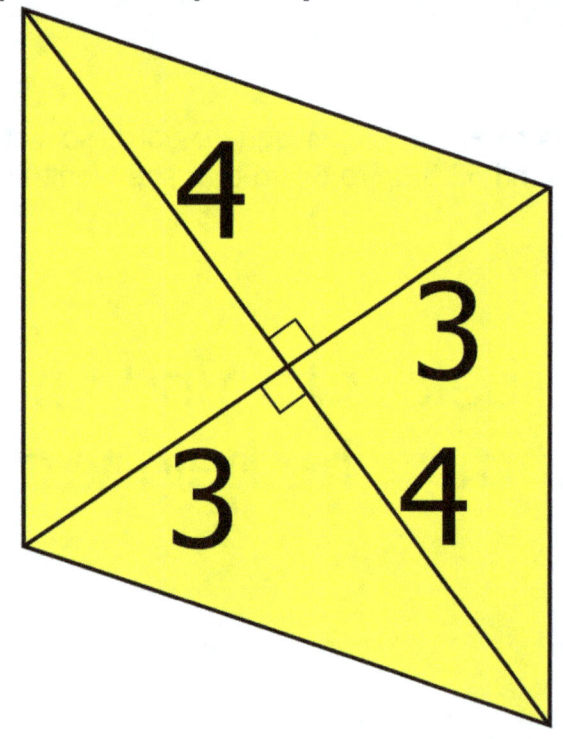

$3 + 3 = 6$
$4 + 4 = 8$
$6 \times 8 = 48$
$48 \div 2 =$ 24 units^2

Find the <u>area</u> of a rhombus whose first diagonal is 4 cm and whose second diagonal is 7 cm.

14 cm^2 (**Solution:** $4 \times 7 = 28$, $28 \div 2 = \underline{14}$)

Finding a Rhombus' Height

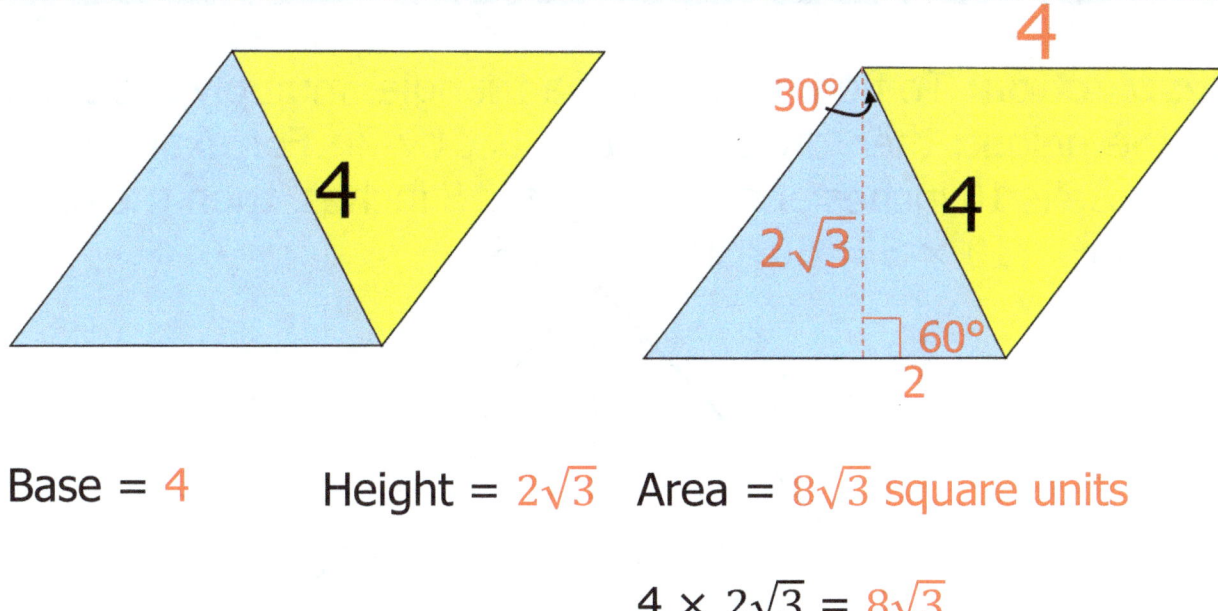

Base = 4   Height = $2\sqrt{3}$ Area = $8\sqrt{3}$ square units

$$4 \times 2\sqrt{3} = 8\sqrt{3}$$

Instruction: The rhombus above has been split into two triangles to help you find its area. The length of the diagonal of this rhombus is 4. The diagonal divided the rhombus into two equilateral triangles, which means that all sides of the triangles have the same length. If you label each side of the triangles as 4, you will see that each side of the rhombus in this example is also 4. (A rhombus has 4 equal sides.) Thus, we have the length of the base of the rhombus (4) to use in the "base × height" formula to find the area. (Note that we cannot use the "$d_1 d_2 \div 2$" formula to find the area of this rhombus because we only know the length of one of its diagonals, and the diagonals of a rhombus are not the same length.)

Now let's use the blue triangle to find the height of the rhombus. Since the triangle is equilateral, all three of its angles must equal 60°. To illustrate the height of the rhombus, draw a perpendicular line (an ***altitude***) that extends from the top of the 60° angle of the blue triangle to the bottom. The perpendicular line splits the top 60° angle in half, creating a 30°-60°-90° triangle inside the larger blue triangle. The hypotenuse of the triangle is also the diagonal of the rhombus. Thus, since we know that the hypotenuse of the 30°-60° right triangle is 4, we also know that the length of the shorter leg will be half that size (2), and the length of the longer leg will be $2\sqrt{3}$. Thus, the height of the rhombus is $2\sqrt{3}$. Now that we know the base and the height, we can multiply them together to find the area of the rhombus. ($4 \times 2\sqrt{3} = 8\sqrt{3}$ square units)

Area of a Triangle: $bh \div 2$

Instruction: To find the area of a triangle, multiply the base by the height; then divide the product by 2. For example, if the base is 10 inches and the height is 5 inches, then the area is 25 in². (10 × 5 = 50, 50 ÷ 2 = 25).

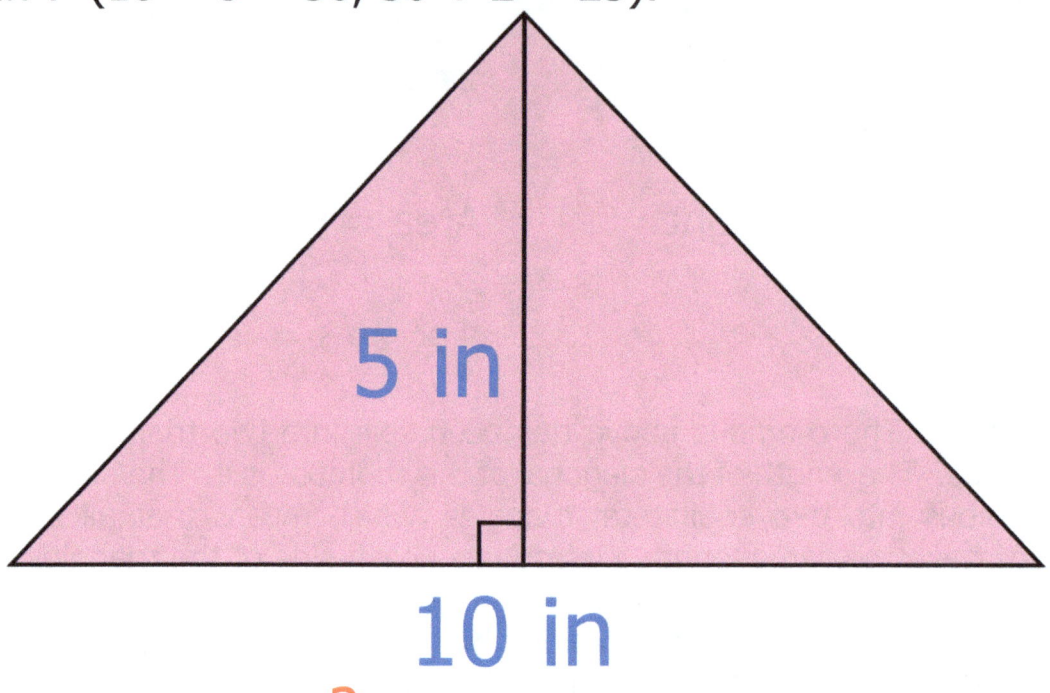

A = 25 in²

If the area of the triangle below is 14, what is its height?

Solution
7 × h ÷ 2 = 14
14 × 2 ÷ 7 = 4

Find the area of an equilateral triangle whose perimeter is 12 in. and whose altitude (height) is $3\sqrt{2}$ in. Use the radical sign in your answer.

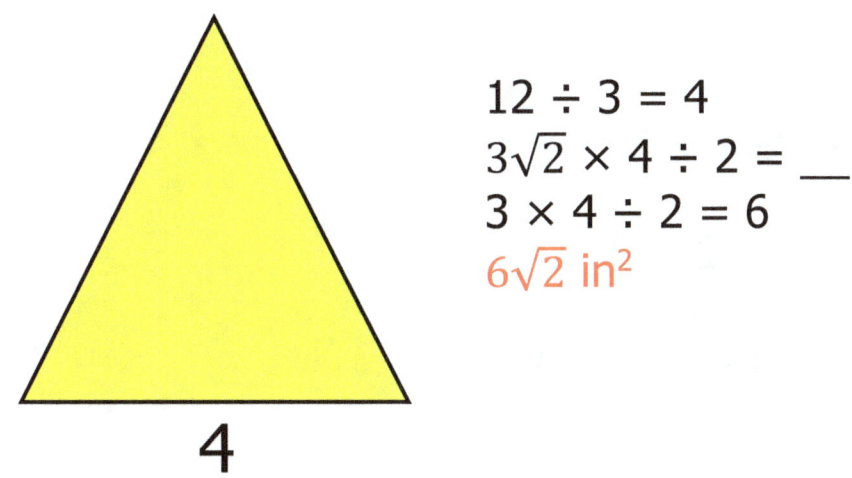

$12 \div 3 = 4$
$3\sqrt{2} \times 4 \div 2 = $ __
$3 \times 4 \div 2 = 6$
$6\sqrt{2}$ in²

Solution: Recall that the formula used to find the area of a triangle is "base × height ÷ 2." Since the perimeter of the given triangle is 12, each side of the equilateral triangle has to be 4. Now that we know that the base is 4, we can use the formula to find the area. Perform the operations without the square root and then add the square root to the answer.

Find the area of the rectangle. Then, use the given value of *x* to find the area of the triangle and of the shaded region.

x = 2

Area of the entire rectangle: 8 × 4 = 32 square units

Area of the triangle: 4 × 2 ÷ 2 = 4 square units

Area of the shaded region: 32 − 4 = 28 square units

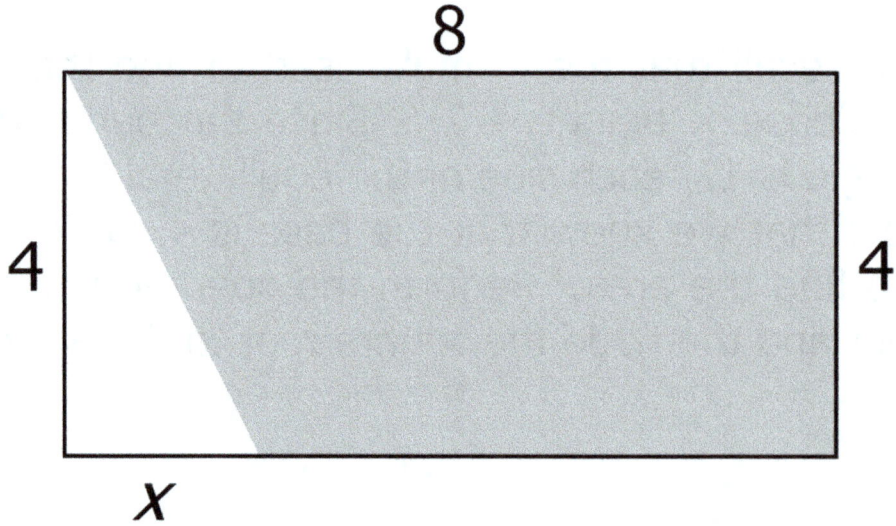

Notice rectangle *EFGH* below. Also notice, triangle *FGI* and triangle *FGH* within the rectangle (point *I* was randomly chosen on $\overline{EH}$). Finally, notice that the height of $\triangle FGI$ is equal to the base of $\triangle FGH$ (see the red dotted lines). This means that the area of both triangles will also be equal. (Recall that the area of a triangle is "base × height ÷ 2.")

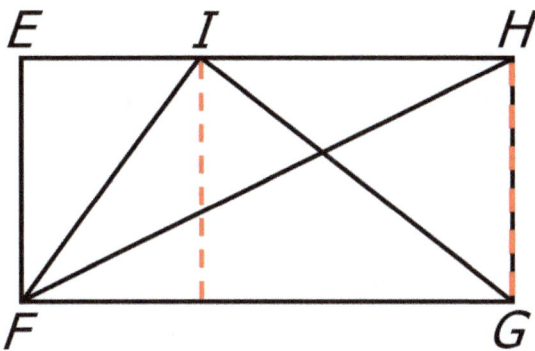

Find the area of a triangle whose sides measure 3, 4, and 5 inches. 6 in²

Hint: You are dealing with a right triangle because 3, 4, and 5 are one of the Pythagorean triples ($3^2 + 4^2 = 5^2$). Since you know that the longest side of a right triangle is the hypotenuse, then the height and base of this particular triangle must be the smaller numbers 3 and 4 (3 × 4 ÷ 2 = 6).

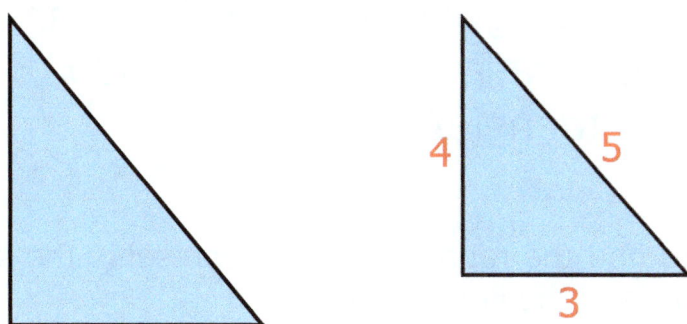

In the figure below, line segments $\overline{DE}$ and $\overline{BC}$ are parallel. Use the information given ($\overline{AD}$ = 6, $\overline{AB}$ = 9, and $\overline{DE}$ = 8) to find the length of $\overline{BC}$ and the area of $\triangle ABC$.

Length of $\overline{BC}$: **12**

Area of $\triangle ABC$: **54 square units**

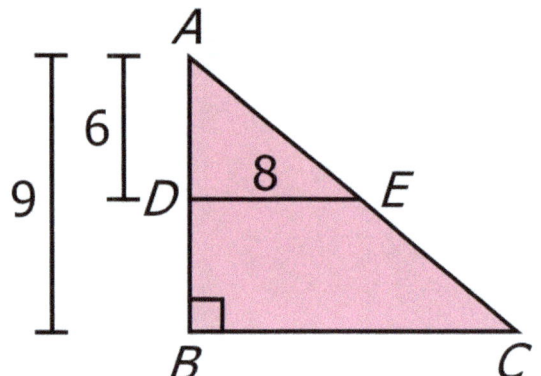

$\dfrac{6}{8} = \dfrac{9}{BC}$

$9 \times 8 = 72, \ 72 \div 6 = 12$
$\overline{BC} = \mathbf{12}$

$12 \times 9 \div 2 = \mathbf{54}$

Instruction: Since the area of a triangle is "base × height ÷ 2" and we need to find the area of triangle *ABC*, then we need to know the length of its base $\overline{BC}$. We know that line segments $\overline{DE}$ and $\overline{BC}$ are parallel, so the corresponding angles of triangles $\triangle ABC$ and $\triangle ADE$ are congruent. And, if the angles are congruent, then the two triangles have to be similar. Recall that the sides of similar triangles are proportional. Thus, we can set up a proportion to find the length of $\overline{BC}$. Record 6 over 8 for the smaller triangle. Then after the equal sign, record 9 over *BC*. (The 9 is recorded directly across from the 6 since it represents the same side of the larger triangle as 6 does of the smaller triangle.) $\dfrac{6}{8} = \dfrac{9}{BC}$

Now, multiply the two known cross numbers (9 and 8) and divide it by the other known number to find the value of $\overline{BC}$ (9 × 8 = 72, 72 ÷ 6 = 12, $\overline{BC}$ = 12). Now that we know that the length of $\overline{BC}$ is 12, we can find the area of $\triangle ABC$.

12 × 9 ÷ 2 = **54 square units**

Since $\triangle ABC$ and $\triangle ADE$ in the figure above are similar, the matching sides have the same ratio. What is the ratio of the smaller triangle to the larger triangle?

$\dfrac{6}{9} = \dfrac{2}{3}$ or **2:3**

Instruction: Reduce 6 over 9 to find the ratio because 6 represents the same side of the smaller triangle as 9 does of the larger triangle.

Finding a Triangle's Height

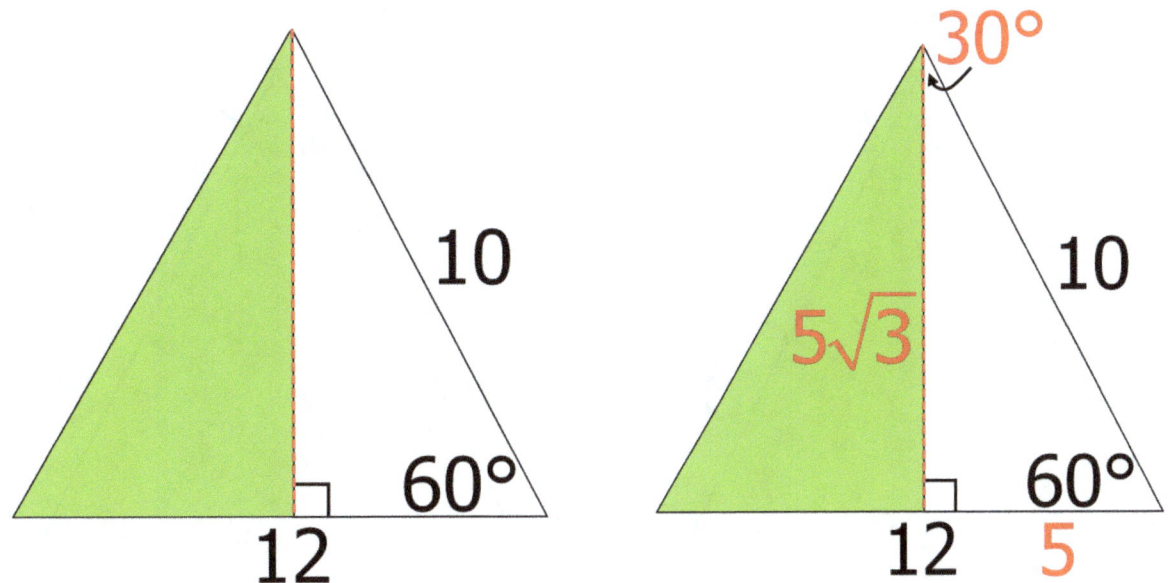

Height of the triangle: 5√3

Area of the triangle: 30√3 units² (Solution: 12 × 5√3 ÷ 2 = 30√3)

Finding the height of a triangle with a 60° angle (Applying what we know about 30°-60°-90° triangles): We are told that one of the angles in the triangle above is 60°, and we know that the angle formed by the altitude is 90°. Thus, the third angle of the white triangle measures 30° because 30 + 60 + 90 = 180. (Recall that 180° is always the sum when you add together all three interior angles of a triangle.) Since the white triangle is a 30°-60°-90° triangle and its hypotenuse is 10, the length of its shorter leg is one-half ten (5), and the length of its longer leg is 5√3. (The height of the original triangle is the same length as the longer leg of the 30°-60°-90° triangle that was formed by the altitude.)

The **area** of the original triangle is base × height ÷ 2, so the area of this triangle is 12 × 5√3 ÷ 2 = 30√3. (Notice that the problem is solved without the √3 but that this square root is recorded in the answer.)

Finding a Triangle's Height

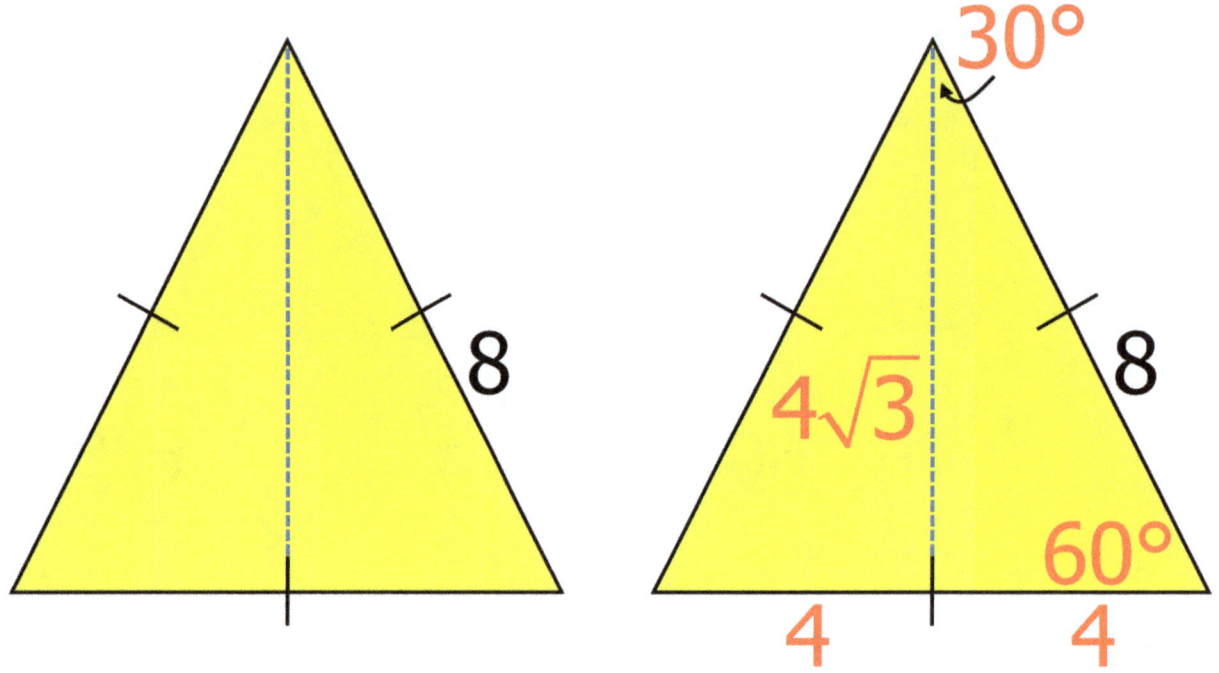

How do we know that the triangle above is equilateral? (All three sides are marked with a single line.)

Finding the height of an equilateral triangle (Applying what we know about 30°-60°-90° triangles): Recall that 180° is the sum of all three angles of any triangle and that each angle of an equilateral triangle measures 60° because 180 ÷ 3 = 60. Also recall that all three sides of an equilateral triangle have the same length.

Dividing an equilateral triangle in half by inserting the altitude (see the dashed line) will form two 30°-60°-90° triangles. Since we know the length of one of the sides of the equilateral triangle, we can use it to find the length of each side of the 30°-60°-90° triangle. Divide the length of the hypotenuse by 2 to find the length of the shorter legs of the 30°-60°-90° triangles (8 ÷ 2 = 4). Finally, multiply the length of the shorter leg by $\sqrt{3}$ to find the length of the longer leg of the right triangles ($4\sqrt{3}$), which is also the height of the equilateral triangle.

Area of a Trapezoid
$$A = h(b_1 + b_2) \div 2$$

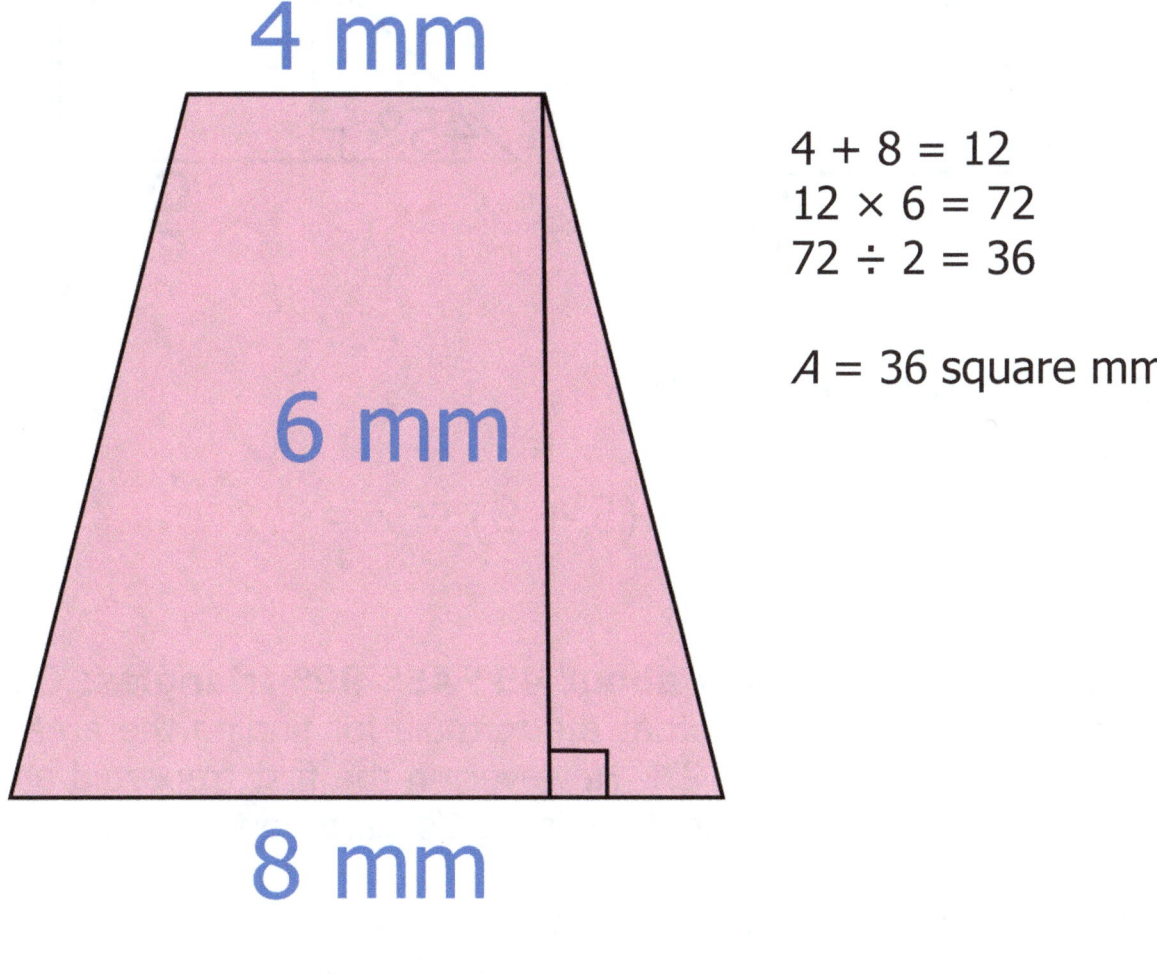

4 mm

6 mm

8 mm

4 + 8 = 12
12 × 6 = 72
72 ÷ 2 = 36

A = 36 square mm

$A = 36 \text{ mm}^2$

Instruction: To find the area of a trapezoid, add base 1 and base 2 (i.e., you add the parallel sides of the trapezoid). Then multiply the sum by the height and divide the product by 2.

Finding a Trapezoid's Height

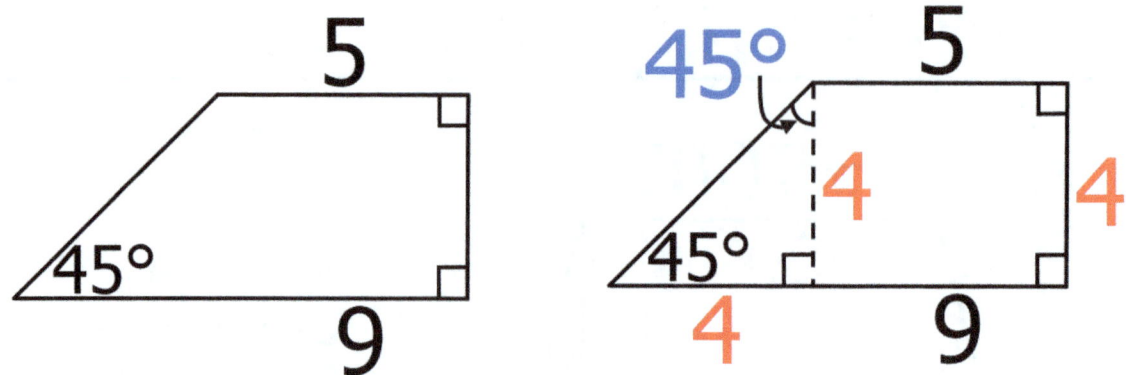

Height of the trapezoid: 4

Area of the trapezoid: 4(5 + 9) ÷ 2 = 28 square units

Applying what we know about 45°-45°-90° triangles: On the previous page, you learned that the formula for finding the area of a trapezoid is "$h(b_1 + b_2) \div 2$." However, in the first trapezoid above, we don't know what the height (h) is. Let's learn how to find it!

First, we draw a perpendicular line from the highest point of the trapezoid to the bottom of the base (see the dashed line). Since the perpendicular line will form a right angle and the other angle measures 45°, the triangle formed by the perpendicular line must be a 45°-45° right triangle because 45 + 45 + 90 = 180. Notice that the top base measures 5 and the bottom base measures 9. Thus, the base of the triangle must be 4 because 4 + 5 = 9. If the bottom leg of the 45°-45° right triangle measures 4, then we know that its other leg (which is also the height of the trapezoid) also measures 4. The area of the trapezoid is: 4(5 + 9) ÷ 2 = **28 square units**.

Area of a Regular Polygon

$$A = pa \div 2$$

Instruction: The **area** of a regular polygon is the perimeter multiplied by the apothem, divided by 2.

In a regular polygon, the segment that extends from the center of the shape to one of its vertices is the **radius** (see the blue line below). In contrast, the **apothem** is the perpendicular segment that extends from the center of the shape to the center of one of its sides (see the red line below).

Since all the sides of a regular polygon have the same length, you can multiply the length of one side by 5 (because a pentagon has 5 sides) to find the perimeter of the pentagon (10 × 5 = 50). Multiply the perimeter by the apothem (50 × 9.2 = 460) and divide the product by 2 to find the area (460 ÷ 2 = 230 units^2).

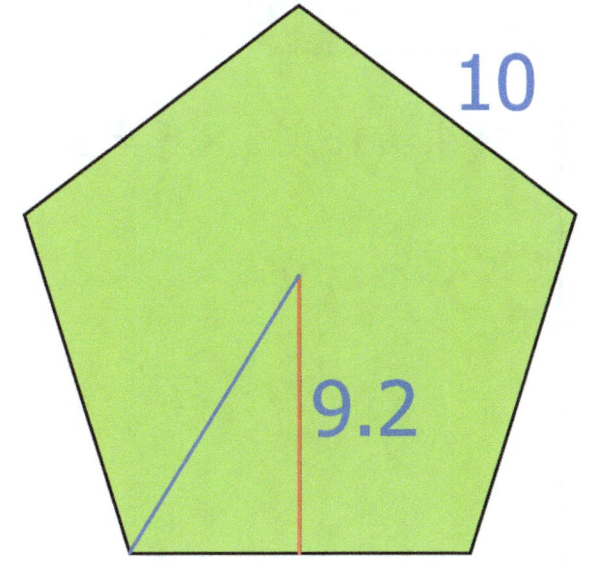

10 × 5 = 50
50 × 9.2 = 460
460 ÷ 2 = 230 units^2

230 units^2

Area and Surface Area

Instruction: To find the area of a square, count the square units or square one of its sides by multiplying it by itself, which is the same as multiplying the base by the height.

To find the surface area of a cube, square one of its sides and multiply the result by 6 (because a cube has 6 square faces).

square s^2
(area)

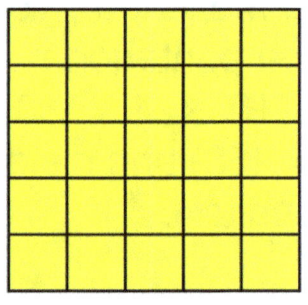

5 cm

$5^2 = 25$

25 cm²

cube $6s^2$
(surface area)

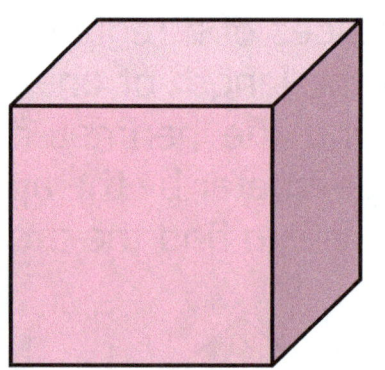

4 cm

$4 \times 4 = 16$, $16 \times 6 = 96$

96 cm²

If the area of a square rug is 49 square feet, what is the length of each of its sides?
7 ft.

Solution: Since you find the area of a square by squaring the length of one of its sides, you will need to do the opposite operation to find the missing length. That is, you find the length of one of its sides by finding the square root of the given area. $\sqrt{49} = 7$ ft

This rectangle is made up of 3 congruent squares. If the area of the rectangle is 48 ft², what is its perimeter? 32 ft

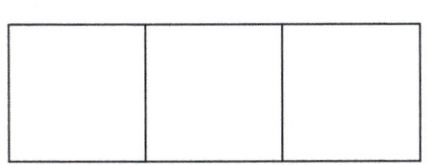 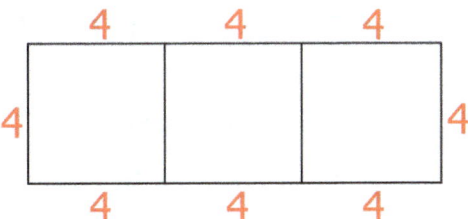

Instruction: Since there are three squares, the area of one square is 16 because 48 ÷ 3 = 16. Recall that to find the area of a square, you square one of its sides. This means that you can find the length of each side of the square by finding the square root of 16 ($\sqrt{16} = 4$). When you insert 4 on the side of each square and add those on the outside together, you find that the perimeter of this rectangle is 32.
    4 + 4 + 4 + 4 + 4 + 4 + 4 + 4 = 32

What is the length of each side of a square whose perimeter is 4 inches? **1 in.**

What is the area of a square whose perimeter is 4 inches? **1 in²** (**Solution:** $1^2 = 1$)

The **perimeter** of the square in the previous problem was 4 inches. If the **area** of this square doubled by increasing the length of its sides, then the new perimeter of the square would be what? **$4\sqrt{2}$ inches**

Solution: Since the new area would be 2 square inches, the length of each side of the square would be $\sqrt{2}$ inches, and the perimeter would be 4 times this length because a square has 4 sides. Thus, the new perimeter would be $4\sqrt{2}$.

What is the circumference of circle P if the area of square ABCD is 16 square units?

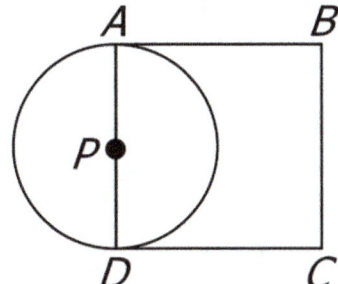

4π (**Solution:** √16 = 4)

Instruction: Since the circumference of a circle is "diameter × π" and the diameter of this circle is also one side of a square that has an area of 16 square units, then the square root of 16 is the diameter (4), and the circumference is 4π.

If the diagonal of a square is 7√2 m, what is the length of each of its sides? 7 m

# What is the area of the square?
7² = 49 m²

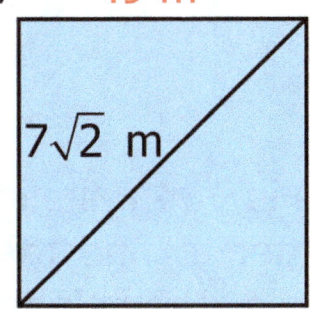
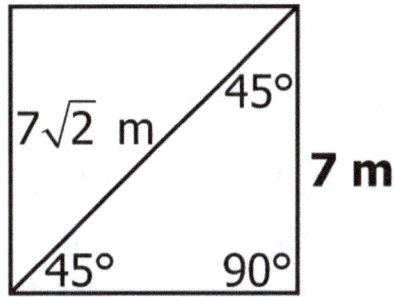

Hint: Recall that one diagonal will divide a square into two 45°-45° right triangles. If the diagonal represents the hypotenuse of the 45°-45° right triangle and you know the diagonal's length, then you can use that length to find the length of the legs of the triangle, which are also the sides of the square.

Find the area of the square in the figure below. 34 square units

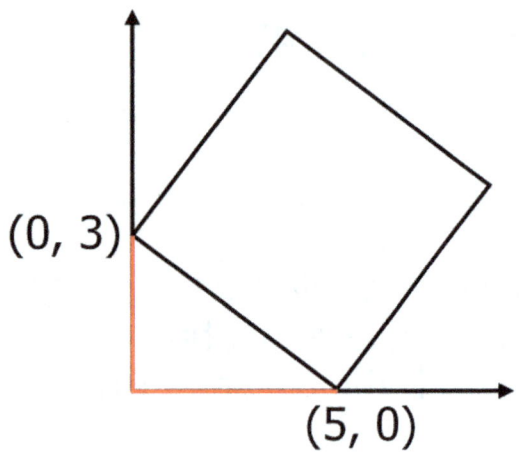

Solution:
$a^2 + b^2 = c^2$
$3^2 + 5^2 = \underline{}$
$9 + 25 = 34, \sqrt{34}$

The area of a square is base times height.
$\sqrt{34} \times \sqrt{34} = 34$

Instruction: Notice the right triangle formed from the square tilting against the *x-* and the *y-*axis on a quadrant of a coordinate plane (see the red lines). Since we know that one leg of the right triangle is 3 units in length and the other leg is 5 units, we can use the Pythagorean Theorem to find the length of the hypotenuse, which is also a side of the square.
   $a^2 + b^2 = c^2$
   $3^2 + 5^2 = \underline{}$
   $9 + 25 = 34, \sqrt{34}$
Now that we know that each side of the square is $\sqrt{34}$ units in length, we can multiply two of its sides together to find the area because the area of a square is the base times the height.
   $\sqrt{34} \times \sqrt{34} = 34$ square units

Find the value of x and the perimeter of the polygon.

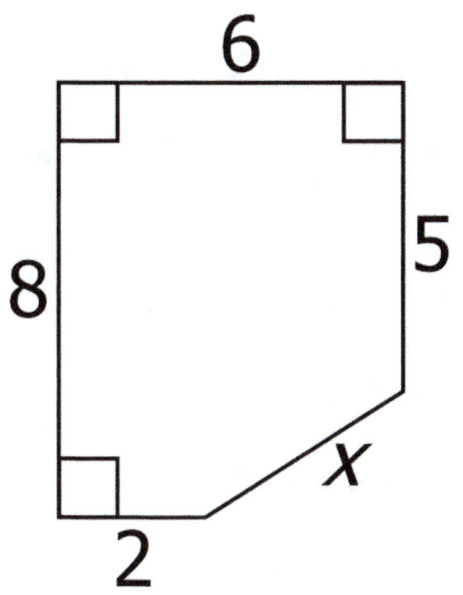 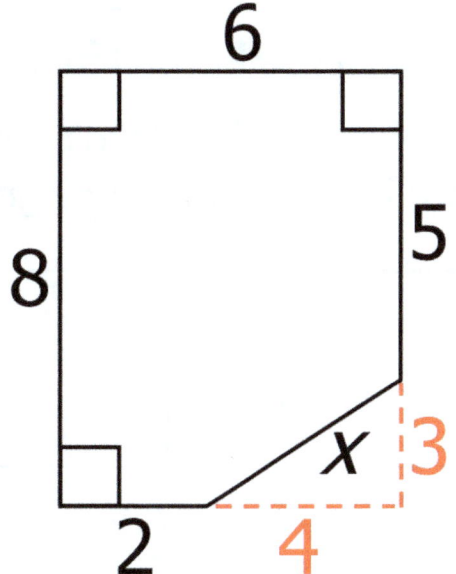

This figure is not drawn to scale.

x = 5

(**Solution:** $6 - 2 = 4$, $8 - 5 = 3$, $4^2 + 3^2 =$ __, $16 + 9 = 25$, $\sqrt{25} = 5$)

Perimeter: 26

(**Solution:** $2 + 8 + 6 + 5 + 5 = 26$)

Instruction: In order to find the perimeter, we have to find the value of x. We can do this by forming a right triangle. We know that the bottom leg of the right triangle is 4 because $6 - 2 = 4$. We also know that the other leg is 3 because $8 - 5 = 3$. Now we can use the Pythagorean theorem to find the value of x since it is also the hypotenuse of the triangle. The value of x is 5 because $4^2 + 3^2 = 25$ and $\sqrt{25} = 5$. Finally, add all the sides together to find the perimeter of the polygon: $2 + 8 + 6 + 5 + 5 = 26$.

In the figure below, the diagonal of the smaller square is the same length as each side of the larger square. Find the area of the smaller square and the area of the larger square. (**Hint:** Don't forget that the diagonal splits the smaller square into two 45°-45° right triangles.)

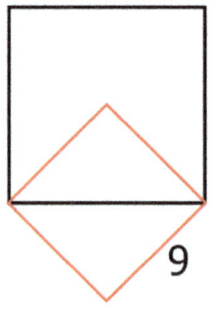
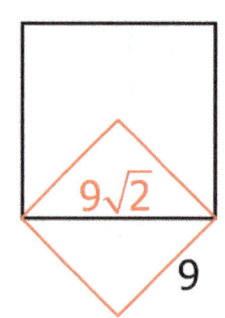

Area of the smaller square:
81 square units
(**Solution:** $9^2 = 81$)

Area of the larger square:
162 square units
Solution: $(9\sqrt{2})(9\sqrt{2}) = 162$

Instruction: Since the diagonal splits the smaller square into two 45°-45° right triangles, the length of the diagonal (and each side of the larger square) must be $9\sqrt{2}$. Square this figure to find the area of the larger square.

The rectangle below is made up of three adjacent squares, and the diagonal of each square is $4\sqrt{2}$ m. Find the area of the shaded part of the rectangle. (**Hint:** Don't forget that the diagonal splits each square into two 45°-45° right triangles.)

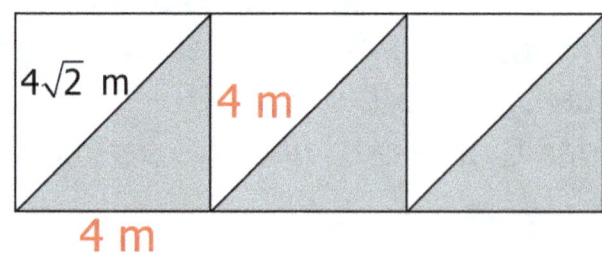

Area of the shaded part: 24 m²

Solution:
$4 \times 4 \div 2 = 8$
$8 \times 3 = 24$

206

In the figure below, square *WXYZ* is inscribed in square *ABCD*. What is the area of the shaded portion of the square if each side of the larger square measures 8 inches? 32 in²

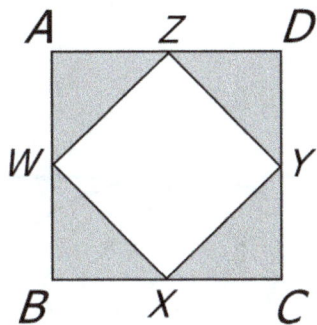

Solution:
8 ÷ 2 = 4
4 × 4 ÷ 2 = 8
8 × 4 = **32**

Instruction: Since the smaller square is inscribed in the larger square, each corner of the smaller square is touching the very middle of the side of the larger square. Notice that the shaded area is made up of four right triangles. First, find the area of one of these triangles, such as $\triangle AWZ$. Both legs of each triangle will measure 4 inches since each leg would be ½ the length of a side of the larger square and 8 ÷ 2 = 4. Thus, the area of the triangle is 4 × 4 ÷ 2 = 8 square units. Since there are 4 triangles, multiply 8 by 4 to find the area of the shaded portion.

In the figure below, the circle is inscribed in a square, and a smaller square is inscribed in the circle. If each side of the smaller square is 5 inches, then the area of the larger square is what? (**Hint:** Don't forget that the diagonal splits the smaller square into two 45°-45° right triangles.)

Area of the larger square: 50 in²

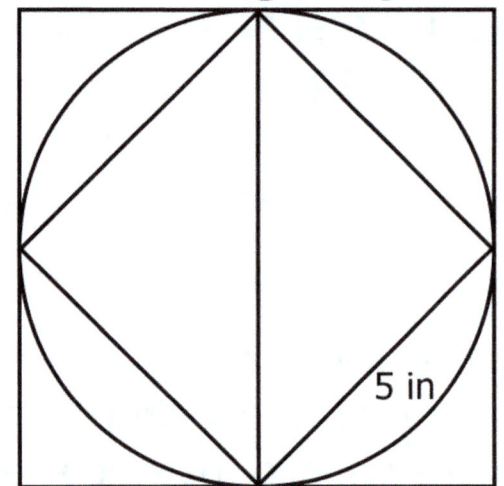
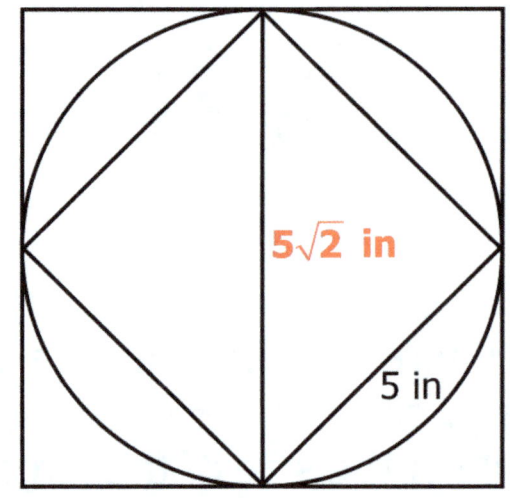

Solution: $(5\sqrt{2})(5\sqrt{2}) = 50$

Instruction: Find the diameter of the circle, which is also the length of the hypotenuse of the 45°-45° right triangle. Since one of the legs of this special triangle is 5 inches, the hypotenuse has to be $5\sqrt{2}$ inches according to the rule for 45°-45° right triangles. Notice that the diameter is the same length as each side of the larger square. Since you square one of the sides of a square to find the area, the area of the larger square is 50 in².

In the coordinate plane to the left, the vertices of a square are plotted on a coordinate plane at (−2, 0), (2, 0), (−2, 4), and (2, 4). What is the area of the square?
16 square units

Solution: When the vertices are plotted on a coordinate plane, you can see that each side of the square is equal to 4 units, and there are 16 square units all together. Thus, the area of the square is 16 square units ($4^2 = 16$).

Teacher instructions: Use an electronic pen to create your own square in the coordinate plane to the right.

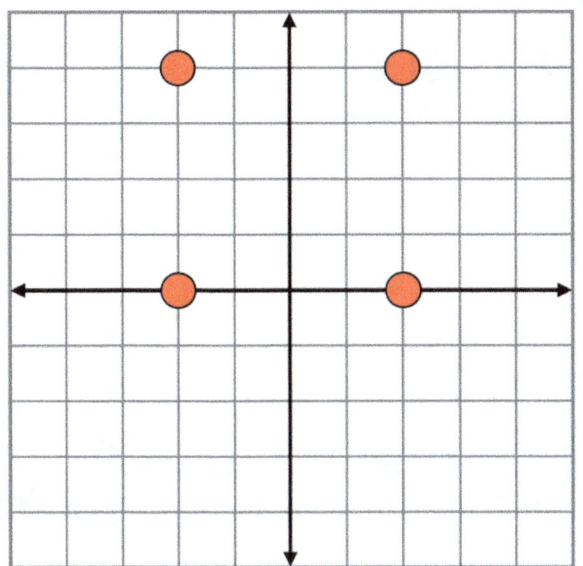
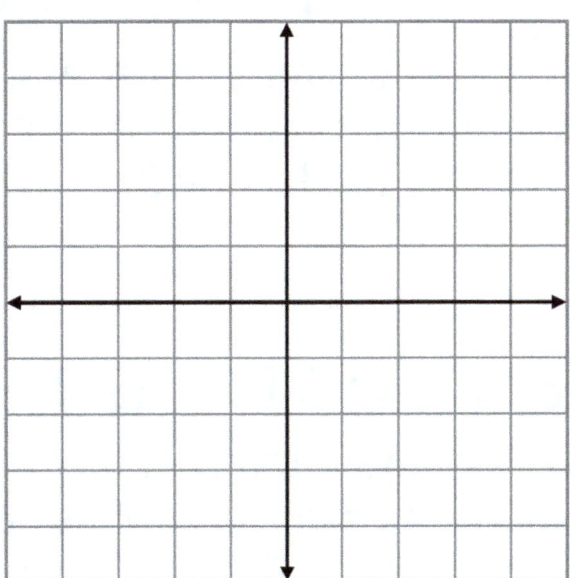

Lateral Surface Area

Instruction: The surface area is the number of square units required to cover the outside of a 3-D figure. The <u>lateral surface area</u>, on the other hand, includes all the surface area other than the two bases.

cylinder: cH

You find the lateral surface area of a <u>cylinder</u> by multiplying the circumference of one of its bases by the height.

right circular cone: $cl \div 2$

To find the lateral surface area of a <u>right circular cone</u>, you multiply the circumference of its base by the slant height and divide the product by 2.

regular pyramid: $pl \div 2$

To find the lateral surface area of a <u>regular pyramid</u>, you multiply the perimeter of its base by the slant height and divide the product by 2.

A cone with a polygon rather than a circle as a base is a pyramid. If it is a regular pyramid, it will have a regular polygon for a base, and the pyramid will be right rather than oblique.

Surface Area (Cylinder)

1. **Lateral surface area (L):** *cH*
   Circumference ≈ 18.84 m
     (or 6π m)
   Height: 10 m

   L: 6π × 10 ≈ 188.4 m²
     (or 60π m²)

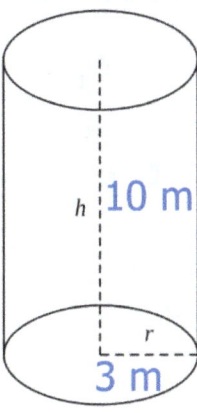

2. Area of one circular **base** ≈ 28.26 m² (or 9π m²)

3. **Surface area (S):** *L + 2B*
   188.4 + 56.52 = 244.92 m² (or 60π + 18π = 78π m²)

Instructions for finding the surface area of the above cylinder

1. **Lateral surface area:** Recall that the surface area is the number of square units required to cover the outside of a 3-D figure. First, you find the lateral surface area (which is the surface area other than the two bases). The formula to find the lateral surface area of a cylinder is "circumference × height." (Remember that you find the circumference by multiplying the diameter by π). The given radius is 3 m. The radius is always ½ the length of the diameter, so the diameter is 6 m, and the circumference is approximately 18.84 m (or 6π m if you don't multiply the circumference by 3.14—the approximate value of pi). After you multiply the circumference by the height, you find that the lateral surface area is approximately 188.4 m² (or 60π m²).

2. **Area of the circular bases:** Next, find the area of one of the circular bases. Recall that the area of a circle is the radius squared times π. The given radius is 3 m, and 3 squared equals 9 (because 3 multiplied by itself equals 9). Finally, 9 × 3.14 = 28.26, so the area of one of the circular bases of this cylinder is approximately 28.26 m² (or 9π m² if you don't multiply the radius squared by 3.14). Multiply this approximate area by 2 because there are two bases (28.26 × 2 = 56.52).

3. **Total surface area:** Finally, find the total surface area by adding the lateral surface area to the area of the two bases. The total surface area is approximately 244.92 m² (or 78π m²).

surface area (right circular cone)

1. **Lateral surface area (L):**
   cl ÷ 2
   Circumference ≈ 25.12 cm
       (or 8π cm)
   Slant height (l): 12 cm

   L: 301.44 ÷ 2 = 150.72 cm²
       or
   8π × 12 = 96π ÷ 2 = 48π cm²
2. **Area of circular base** ≈
   50.24 cm² (or 16π cm²)

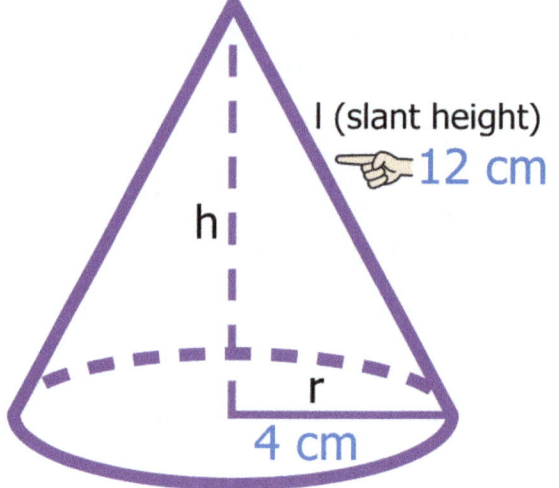

3. **Surface area (S):** *L + B*
   150.72 + 50.24 = 200.96 cm²
   (or 48π + 16π = 64π cm²)

Instructions for finding the surface area of the right circular cone

1. **Lateral surface area:** First you find the lateral surface area (which is the surface area other than the circular base). The formula to find the lateral surface area of a cone is "circumference × slant height ÷ 2." (Remember that you find the circumference by multiplying the diameter by π). Since the radius of this cone is 4 cm, the diameter is 8 cm, and the circumference is approximately 25.12 cm (or 8π cm if you don't multiply the diameter by 3.14—the approximate value of pi). After you multiply the circumference by the slant height and divide the product by 2, you find that the approximate lateral surface area is 150.72 cm² (or 48π cm²).
2. **Area of the circular base:** Next, you find the area of the circular base. Recall that the area of a circle is the radius squared times π. The radius of this circle is 4 cm. When you square 4 by multiplying it by itself, you get 16. Multiply 16 by 3.14 (the approximate value of pi) and you find that the area of the circular base is approximately 50.24 cm² (or 16π square units if you don't multiply the radius squared by 3.14).
3. **Total surface area:** Finally, find the total surface area by adding the lateral surface area to the area of the circular base. The total surface area is approximately 200.96 cm² (or 64π cm²).

Surface Area (Regular Pyramid)

1. **Lateral surface area (L):**
   6 × 14 ÷ 2 = 42;
   42 × 5 = 210 ft²
   (or you could use the formula
   pl ÷ 2: 30 × 14 ÷ 2 = 210)

2. Area of the **base**: *pa* ÷ 2
   Perimeter: 30 ft

   Apothem: 3.4 ft
   30 × 3.4 ÷ 2 = 51 ft²

3. **Surface area (S)**: *L* + *B*
   210 + 51 = 261 ft²

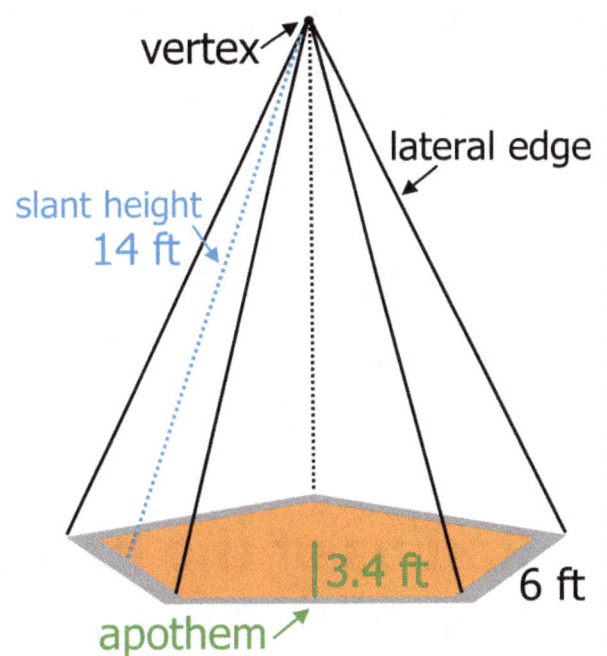

Notice that 3.4 ft is the apothem and 14 ft is the slant height.

Instructions for finding the surface area of a regular pyramid

1. **Lateral surface area:** Recall that the surface area is the number of square units required to cover the outside of a 3-D figure. You first find the area of the lateral faces (i.e., the triangular faces). The area of a triangle is "base × height ÷ 2." (Note that the **slant height** is also the height of one of the pyramid's triangular lateral faces.) The area of one of the lateral faces is 42 ft² because 14 × 6 ÷ 2 = 42. Multiply 42 by 5 because there are 5 lateral faces (we know this because the base has 5 sides). The lateral surface area is 210 ft² because 42 × 5 = 210. (The formula "*pl* ÷ 2 = lateral surface area" could also be used.)
2. **Area of the base:** Next, you find the area of the regular polygon base by multiplying the perimeter by the apothem and dividing the product by 2 (*pa* ÷ 2).
3. **Total surface area:** Finally, find the total surface area by adding the lateral surface area to the area of the base. The total surface area is <u>261 ft²</u>.

Instruction: When you know the radius of the base and the height of a right circular cone or regular pyramid, you can use the formula below to find the slant height. This formula tells us to square the radius of the base and the height of the cone or pyramid, to add the two numbers together, and then to find the square root of the sum.

$$\sqrt{radius^2 + height^2} = \text{slant height}$$

Use the information given below to find the slant height and the surface area of a right circular cone. Round decimals to the nearest hundredths.

Radius of the base: 3 cm
Height of the cone: 6 cm

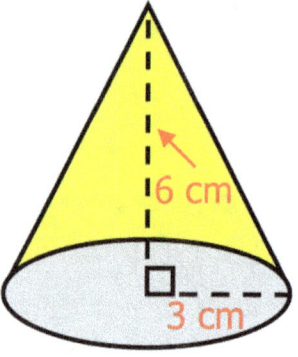

Slant height: 6.71 (**Solution:** $\sqrt{3^2 + 6^2} = \sqrt{9 + 36} = \sqrt{45} = 6.71$)

Now find the surface area of the cone above. Recall that to find the **surface area** of a right circular cone or regular pyramid, we add the lateral surface area to the area of the base (*L + B*). Since the base is a circle and the radius of the circle is 3 cm, the **area of the circular base** is 9π cm² (the area of a circle is the radius squared times π).

The formula to find the **lateral surface area** is "*cl* ÷ 2" (*c* stands for *circumference* and *l* for *slant height*. Multiply the diameter by π to find the circumference. Since the radius of the circular base we are working with is 3 cm, the diameter is 6 cm, and the circumference is 6π cm. Thus, the lateral surface area is 20.13π cm² or 63.24 cm² ($6\pi \times 6.71 \div 2 = 63.24$).

Area of the circular base: $3^2 = 9$, $9 \times \pi = 9\pi$ cm²
Lateral surface area: *cl* ÷ 2, $6\pi \times 6.71 \div 2 = 20.13\pi$ or 63.24 cm²
Surface area: L + B, $20.13\pi + 9\pi = 29.13\pi$ cm² or 91.51 cm²

Notes

The blue square below represents the base of the regular pyramid next to it. Notice that the diagonal of the square measures $6\sqrt{2}$ cm. If the surface area of the pyramid is 96 cm², what is the slant height of the pyramid?

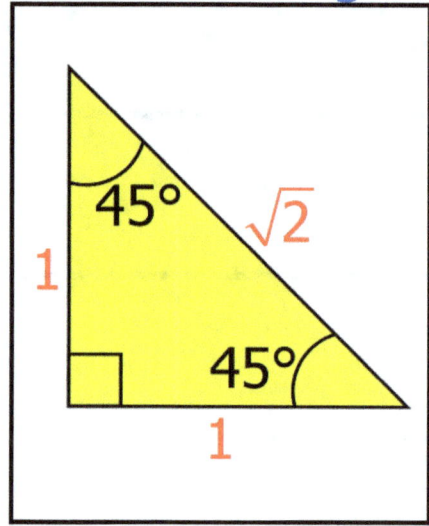
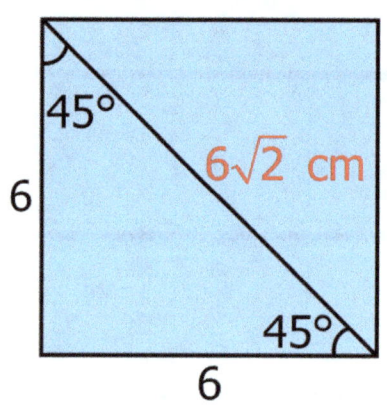
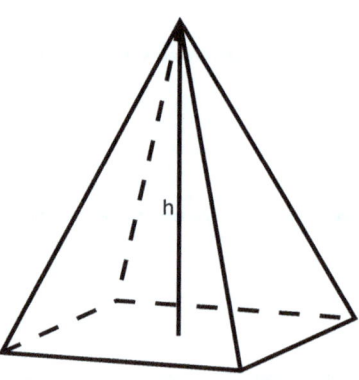

Instruction: Recall that one diagonal will divide a square into two 45°-45° right triangles. In the triangle illustrated in the box above, we note that if the sides of the triangle are 1, then the hypotenuse is $\sqrt{2}$. However, since we are told that the diagonal (or the hypotenuse) in this problem measures $6\sqrt{2}$ instead of $\sqrt{2}$, then we also know that the sides of the square in this problem measure 6 and not 1. Now we can find the area of the square.

Base of the regular pyramid (the area of the square)
   6^2 = **36 cm²**

Lateral surface area (L) of the regular pyramid
The problem tells us that the surface area of the pyramid is 96 cm².
Thus, we could write:
   L + B = Surface area
   L + 36 = 96 cm²
You can then find the lateral surface area by subtracting 36 from 96.
   96 − 36 = **60** (The lateral surface area is 60 cm².)

Slant height of the regular pyramid
Recall that the formula used to find the lateral surface area of a regular pyramid is "$pl \div 2$," where p stands for the perimeter of the base and l stands for the slant height.

$pl \div 2$ = lateral surface area
$24 \times l \div 2 = 60$
(The perimeter of the square is 24 because $6 \times 4 = 24$. Work backwards using the opposite operations to find the slant height: $60 \times 2 \div 24 = 5$).
$l =$ 5 cm (The slant height is 5 cm.)

To find the altitude (the height) of the pyramid on the previous page, insert the slant height you found in the previous problem into the following formula:

$h^2 +$ (1 side of the square $\div$ 2)$^2 = l^2$
(The variable h represents the height, while l represents the slant height.)
$h^2 + (6 \div 2)^2 = 5^2$
$h^2 + (3)^2 = 5^2$
$h^2 + 9 = 25$
$h^2 = 16, \sqrt{16} = 4$
$h =$ 4 cm

The circular base of a certain right cone has a perimeter of 10π cm. If the surface area is 60π cm², what is the slant height and the height of the cone?

Instruction: Recall that the perimeter of a circle is its circumference. Since we find the circumference of a circle by multiplying its diameter by π, we know that the diameter of a circle that has a circumference of 10π cm is 10 and that its radius is 5.

Base of the right circular cone (the area of the circle)
$r^2 \times \pi$
$5^2 =$ **25π cm²**

Lateral surface area (L) of the right circular cone
  L + B = Surface area
  L + 25π = 60π cm²
  (60π − 25π = 35π — The lateral surface area is 35π.)

Slant height of the cone
  ***cl* ÷ 2 = lateral surface area**
(*C* stands for the circumference of the circle and *l* for the cone's slant height.)
  10π × *l* ÷ 2 = 35π
  35π × 2 ÷ 10π = 7
  (*l* = 7 — The slant height is 7 cm.)

To find the height (the altitude) of the cone, insert the slant height you found in the previous problem into the following formula:
$h^2 + r^2 = l^2$
$h^2 + 5^2 = 7^2$ $h^2 + 25 = 49$
$h = \sqrt{24}$ cm

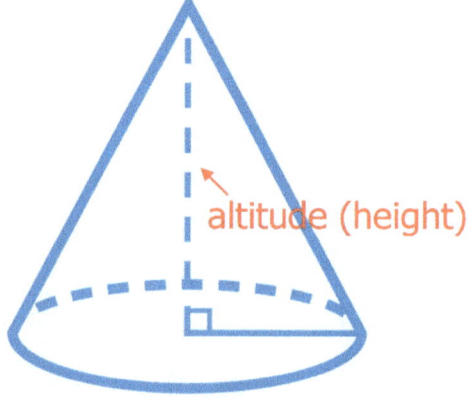
altitude (height)

218

Surface Area (Prism)

Instruction: Find the area of each rectangle by multiplying the base by the height. Then add all the products together.

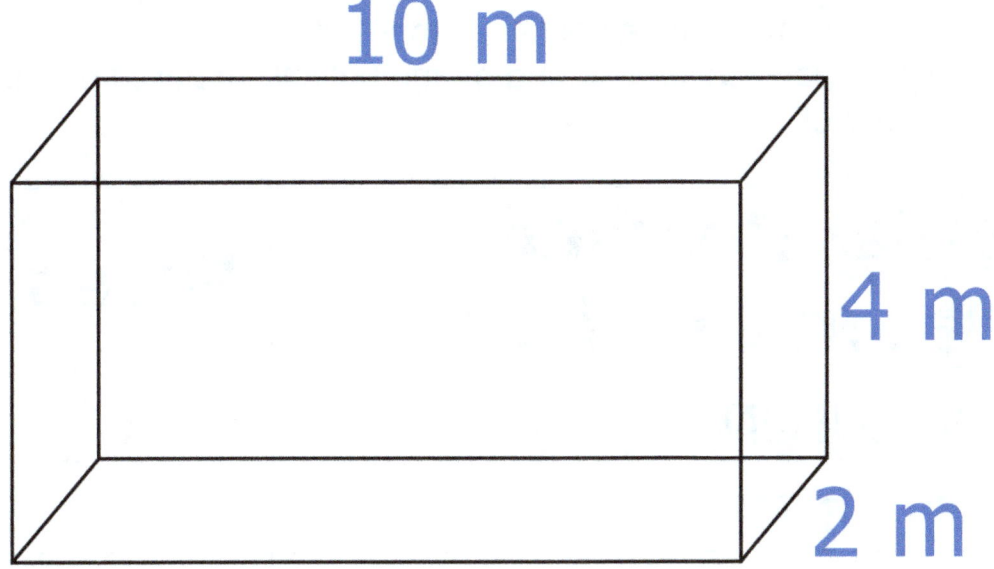

2 × 4 = 8, 10 × 4 = 40, 10 × 2 = 20
2 × 4 = 8, 10 × 4 = 40, 10 × 2 = 20

8 + 8 + 40 + 40 + 20 + 20 = 136

136 m^2

Platonic Solids (Surface Area)

Instruction: We've already learned that a **regular polygon** (such as a square) is both equilateral and equiangular. Now we will learn about **regular solid figures**. Plato discovered that there were only five regular polyhedra (also known as *Platonic solids* after Plato). Each face of a regular polyhedron is a regular polygon. Also, all the faces are exactly alike (they are congruent), and the same number of faces meet at each vertex. The surface area of a regular polyhedron is found by multiplying the area of one face by the total number of faces.

Regular Polyhedra (Platonic solids)	Faces
1. Tetrahedron	4
2. Hexahedron	6
3. Octahedron	8
4. Dodecahedron	12
5. Icosahedron	20

If a face of a regular tetrahedron covers 7 square inches, what is its surface area?
28 in^2

Solution: Since you know that the faces of a regular polyhedron are congruent and are told that the area of one face is 7 square inches, simply multiply the area of one face by 4 because a tetrahedron has 4 faces.

Chapter 6

Volume, Trigonometry, Sets, and More

(Suggested Grades: 8th and 10th)

Note to teachers: When studying chapter 6, please also take time to review pages 103, 105, 106, and 184 with your class, as they contain information students need to know when completing certain classwork/homework and test problems for this chapter.

Teacher instructions: Using *70 Times 7 Math: Electronic Textbook for Teachers (Geometry for Middle and High School Students),* ask students to identify any missing answers for you to write on the screen. Please note that since the answers are provided in student textbooks, they should have them closed during this time. Student textbooks can also be used as a key for the teacher's benefit.

Volume Formulas

Instruction: The formula used to find the volume of a <u>rectangular prism</u> (also called a *cuboid*) is length times width times height. The formula used to find the volume of a <u>prism or a cylinder</u> is the area of one base times the height. The formula used to find the volume of a <u>regular pyramid or a right circular cone</u> is the area of one base times the height, divided by 3.

rectangular prism (cuboid):
$l \times w \times h$

prism and cylinder:
area of 1 base × h

pyramid and cone:
area of base × height ÷ 3

Volume Formulas

Instruction: To find the volume of a <u>cube</u>, you cube the length of one of its edges by multiplying the length by itself three times.

The volume of a <u>sphere</u> is "4 ÷ 3 times π (pi) times the radius cubed."

cube: e^3

sphere: $\frac{4}{3}\pi r^3$

Volume of a Rectangular Prism (Cuboid)
l × w × h

Instruction: To find the volume of a rectangular prism (also called a *cuboid*), multiply the length by the width by the height.

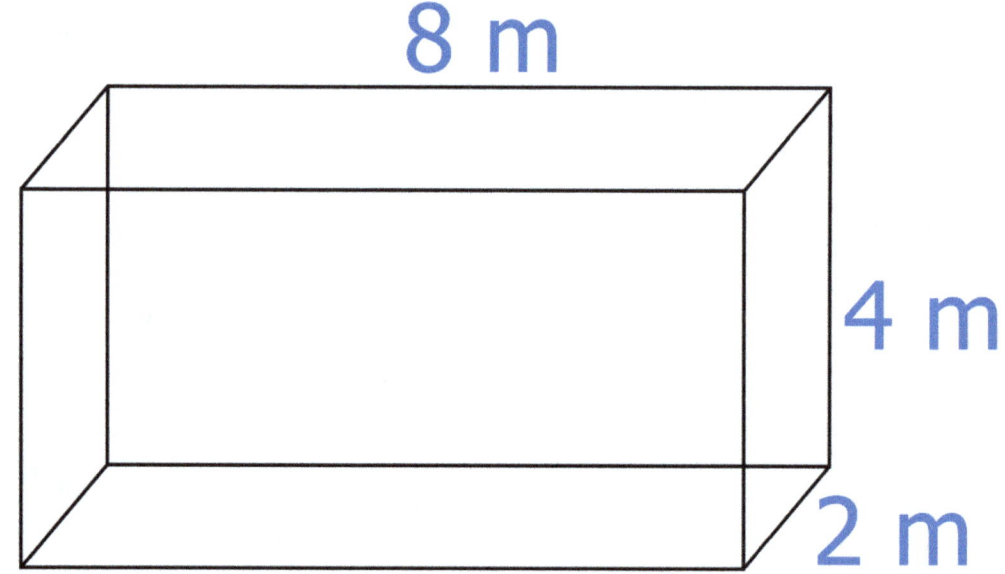

8 × 2 × 4 = 64 m³

Miranda cut a square from all four corners of a rectangular sheet of paper and then folded up the sides to make a lidless box. If the rectangle was 16 in. by 10 in. and the perimeter of each square she cut was 12 inches, what is the volume of her box?

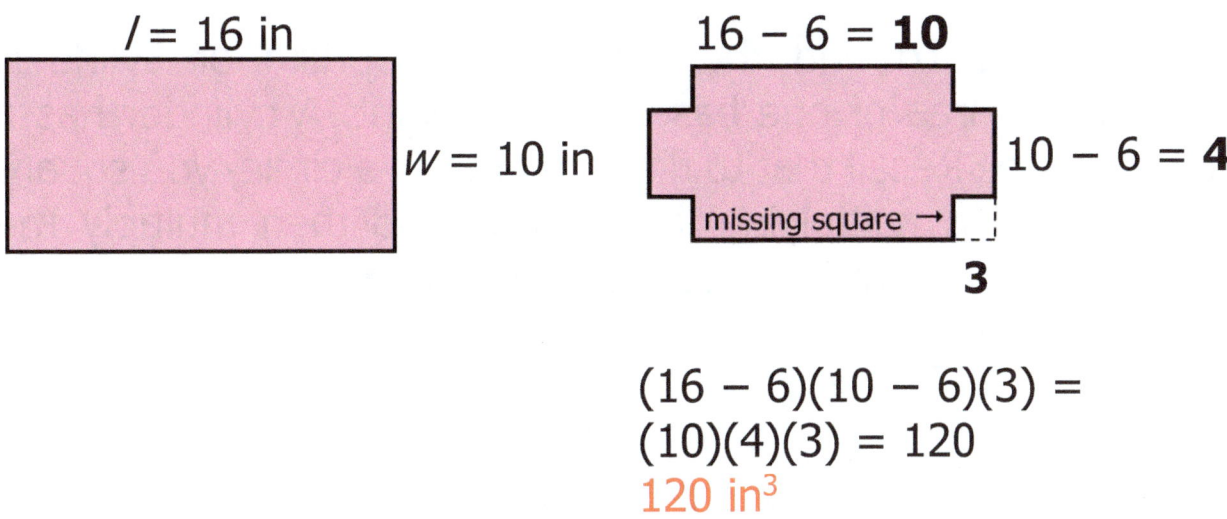

$(16 - 6)(10 - 6)(3) =$
$(10)(4)(3) = 120$
120 in³

Instruction: Recall that you find the volume of a rectangular prism (or a box figure) by multiplying the length by the width by the height ($l \times w \times h$). If the perimeter of one of the squares she cut from a corner was 12 inches, then each side of that square is equal to 3 inches (because $3 + 3 + 3 + 3 = 12$). What this means is that the height of the box also equals **3** inches (see the illustration above), the length equals **10** (because you subtract 3 from each corner of the length and $16 - 6 = 10$), and the width is equal to **4** (because you subtract 3 from each corner of the width and $10 - 6 = 4$). Thus, the volume of Miranda's box is 120 in³ because $(10)(4)(3) = 120$.

Volume of a Prism or Cylinder

area of 1 base × height

Instruction: To find the volume of a prism or cylinder, multiply the area of one base by the height. A cylinder has a circular base. Recall that to find the area of a circle, you square the radius (by multiplying it by itself) and then multiply the result by π (pi), which is approximately 3.14.

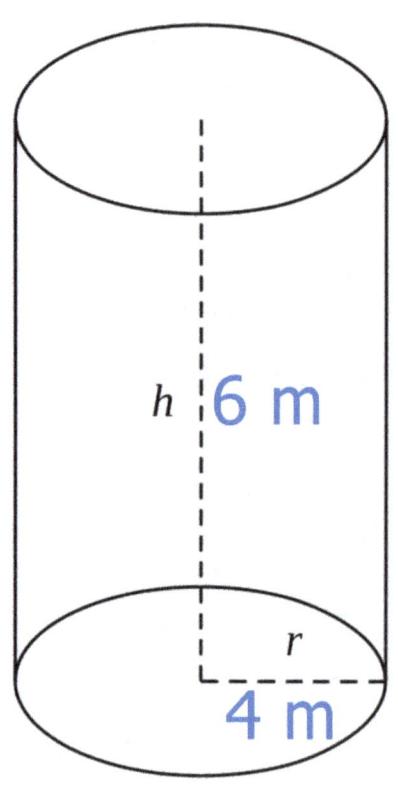

$A = 301.44 \text{ m}^3$

Formula: $\pi r^2 \times h$

$4^2 = 16$, $16 \times 3.14 = 50.24$

$50.24 \times 6 = 301.44 \text{ m}^3$
(or $96\pi \text{ m}^3$)

Record and solve an equation to find the height of a cylinder. The area of one of its bases is 30.18 m², and its volume is 271.62 m³.

(**Hint:** Before writing the equation, ask yourself, "What is the formula for finding the volume of a cylinder?")

Record the equation: __ × 30.18 = 271.62

Height: 9 m
(**Solution:** 271.62 ÷ 30.18 = 9 m)

Miranda's daddy needs to buy enough sand to fill the new sandbox he built her. If the sandbox is 7 feet square and 2 feet deep, how much sand does he need?

98 ft.³
(**Solution:** 7² × 2 = 98)

Mandi wants to use the rectangle below to form a cylinder. If she rolls the rectangle so that the left side touches the right side without overlapping it, then the volume of the cylinder would be what? (**Answer:** 150)

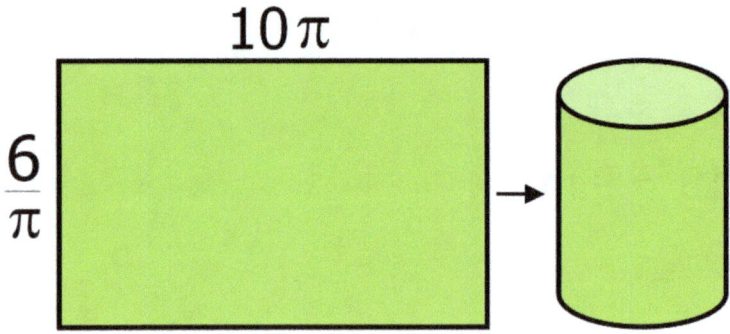

Area of one of the circular bases: 25π square units
($5^2 \times \pi = 25\pi$)

Volume of the cylinder: 150 cubic units
$$\frac{25\pi}{1} \times \frac{6}{\pi} = \frac{150\pi}{\pi} = \mathbf{150}$$

Instruction: The formula used to find the volume of a cylinder is "**area of one base × height**." We are given the height ($6/\pi$), but we need to find the area of one of its bases. Since the circumference of the circle will be 10π, the diameter is 10 and the radius is 5. The area of a circle is "$\boldsymbol{r^2 \times \pi}$," so the area of one of the bases of the cylinder is 25π square units ($5^2 \times \pi = 25\pi$). After you find the area of one of the circular bases, multiply it by the given height to find the volume.

Volume (Pyramid or Cone)
area of base × height ÷ 3

Instruction: To find the volume of a pyramid or cone, find the area of one base; multiply the area by the height; and then divide by 3.

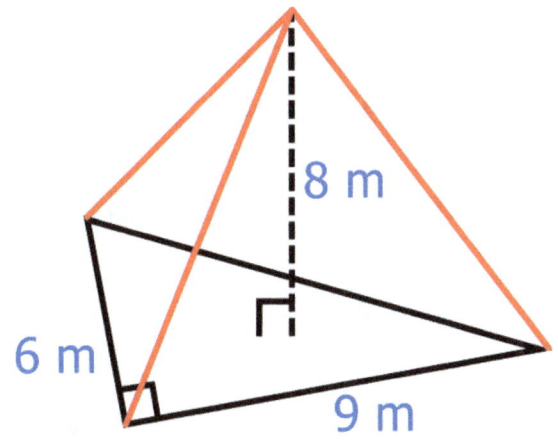

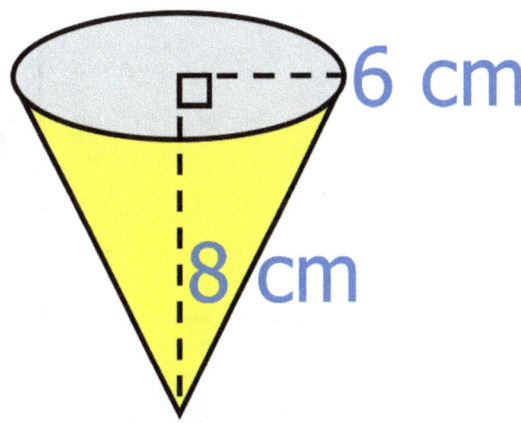

$6 \times 9 \div 2 = 27$
$27 \times 8 \div 3 = 72 \text{ m}^3$

$\pi 6^2 = 113.04$
$113.04 \times 8 \div 3 = 301.44 \text{ cm}^3$
(or $96\pi \text{ cm}^3$)

Instruction: Find the area of the right triangular base. Recall that the area of a triangle is "base × height ÷ 2." The area of this triangle is 27 m because $6 \times 9 \div 2 = 27$. The area of the triangle is then multiplied by the height of the pyramid and divided by 3 ($27 \times 8 \div 3 = 72$.) The volume of the triangular pyramid is 72 m³.

Instruction: A cone has a circle base. Recall that the formula used to find the area of a circle is "$r^2 \times \pi$" (you square the radius by multiplying it by itself and then multiply the result by pi or 3.14).

The figure below shows the base of a certain **cone**. If the altitude (height) of the cone is 5 meters, what is the volume of the cone?

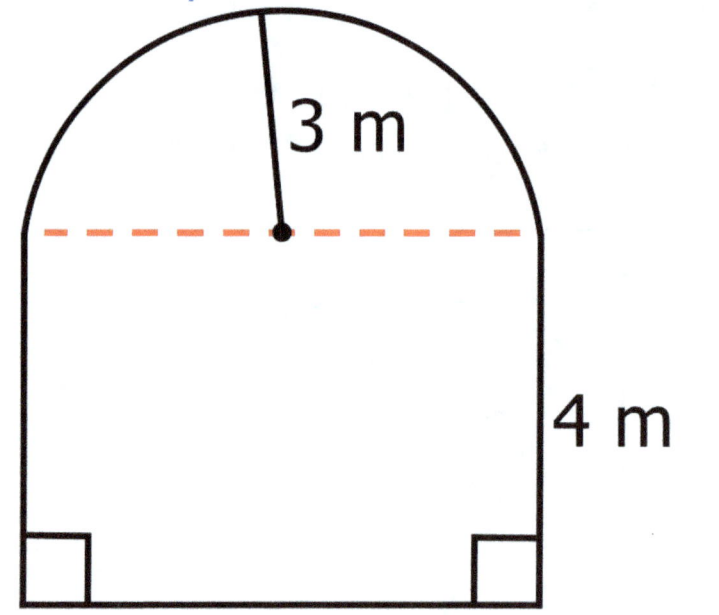

Solution:
area of base × height ÷ 3

4 × 6 = **24**

$3^2 \times \pi = 9\pi$
$9\pi \div 2 = \mathbf{4.5\pi}$

24 + 4.5π = **38.13**

38.13 × 5 ÷ 3 = **63.55 m³**

Instruction: The red dotted line represents the diameter of the half circle as well as one side of the rectangle. Since the diameter is twice the size of the radius (which is 3), the diameter (and one side of the rectangle) is 6. Now we can figure out the area of the rectangle.

Area of the rectangle: *bh* (4 × 6 = 24 m²)

Area of the half circle: We already learned that the formula for finding the area of a circle is "$r^2 \times \pi$." Since we are looking for the area of a half circle, we will need to divide the answer by 2.
   $3^2 \times \pi = 9\pi$, $9\pi \div 2 = 4.5\pi$ m²

Area of the base: 24 + 4.5π = **38.13 m²**

The formula for finding the volume of a cone or pyramid:
   area of base × height ÷ 3
   38.13 × 5 ÷ 3 = **63.55 m³**

Volume of a Cube
e^3

Instruction: To find the volume of a cube, you cube the length of one of its edges.

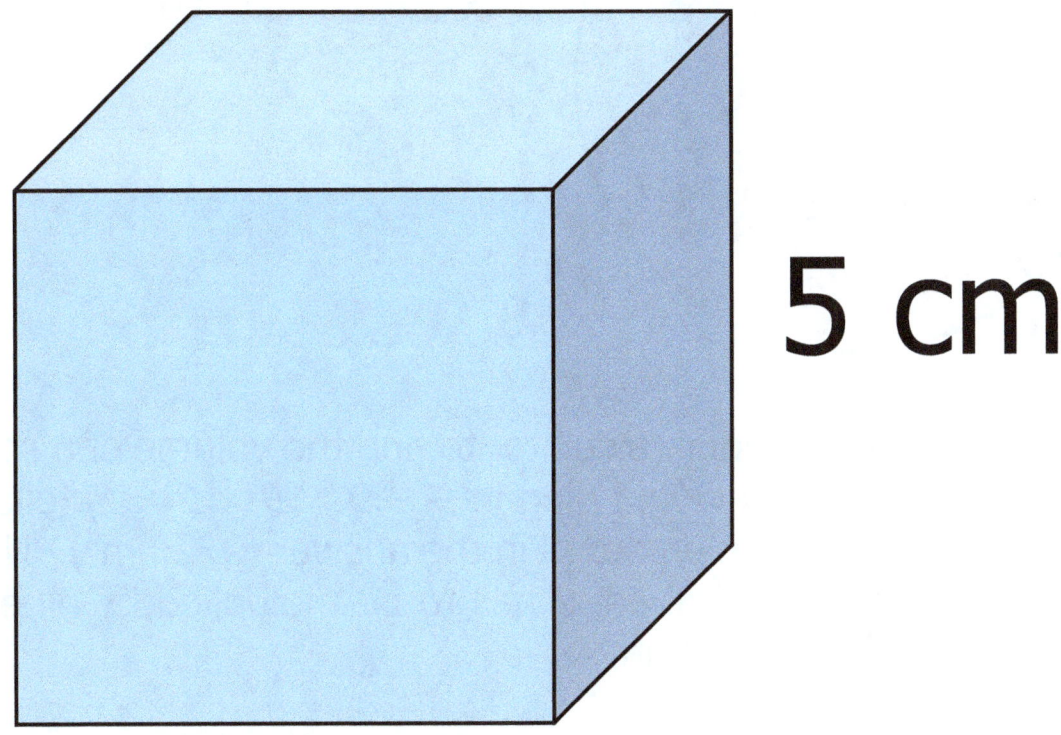

5 cm

5^3 = 125 cm³

5 × 5 × 5 = 125

Volume of a **Sphere**: $\frac{4}{3}\pi r^3$

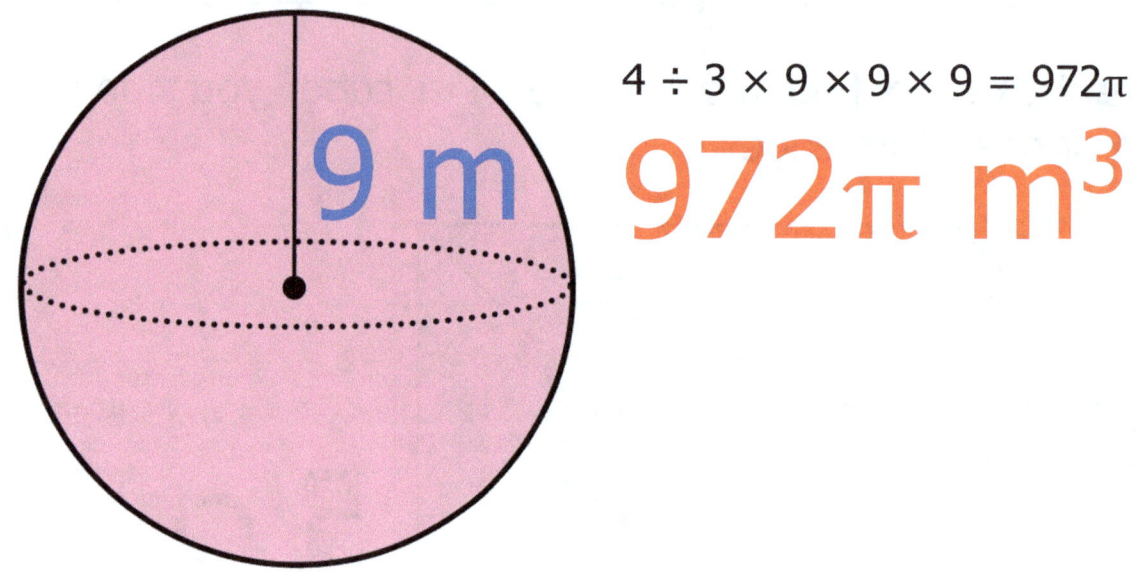

$4 \div 3 \times 9 \times 9 \times 9 = 972\pi$

$972\pi \text{ m}^3$

Instruction: To use your calculator to find the volume of a sphere with a radius of 9, you would plug in *4 ÷ 3 × 9 × 9 × 9*, followed by the equal sign. Then record pi in the answer (972π m³). Notice that since the radius is cubed, 9 has to be multiplied by nine and then multiplied by nine again.

If the volume of a sphere is equal to its surface area, what is the radius of the sphere? 3

Instruction: Recall that the formula used to find the surface area of a sphere is $4\pi r^2$, while the formula used to find the volume of a sphere is $\frac{4}{3}\pi r^3$.

Make one formula equal the other.
$$\frac{4}{3}\pi r^3 = 4\pi r^2$$

Cancel 4 and π on both sides of the equal sign.
$$\frac{1}{3}r^3 = r^2$$

Move 3 in the fraction to the opposite side of the equal sign by multiplying it by r^2.
$$r^3 = 3r^2$$

Move r^2 to the opposite side of the equation by dividing it by r^3. You can use 1 for r on your calculator. When we do this, we can see that the radius equals 3.
$$r = 3$$

Measurements (Length)

25 millimeters (mm) ≈ 1 inch

2½ centimeters (cm) ≈ 1 inch

Instruction: Point out 1 mm, 1 cm, and 1 inch on the tape measure. It takes approximately 25 millimeters (or 2½ centimeters) to equal one inch. The wavy equal sign means "almost equal to." Is the millimeter, the centimeter, or the inch longest? Which is shortest?

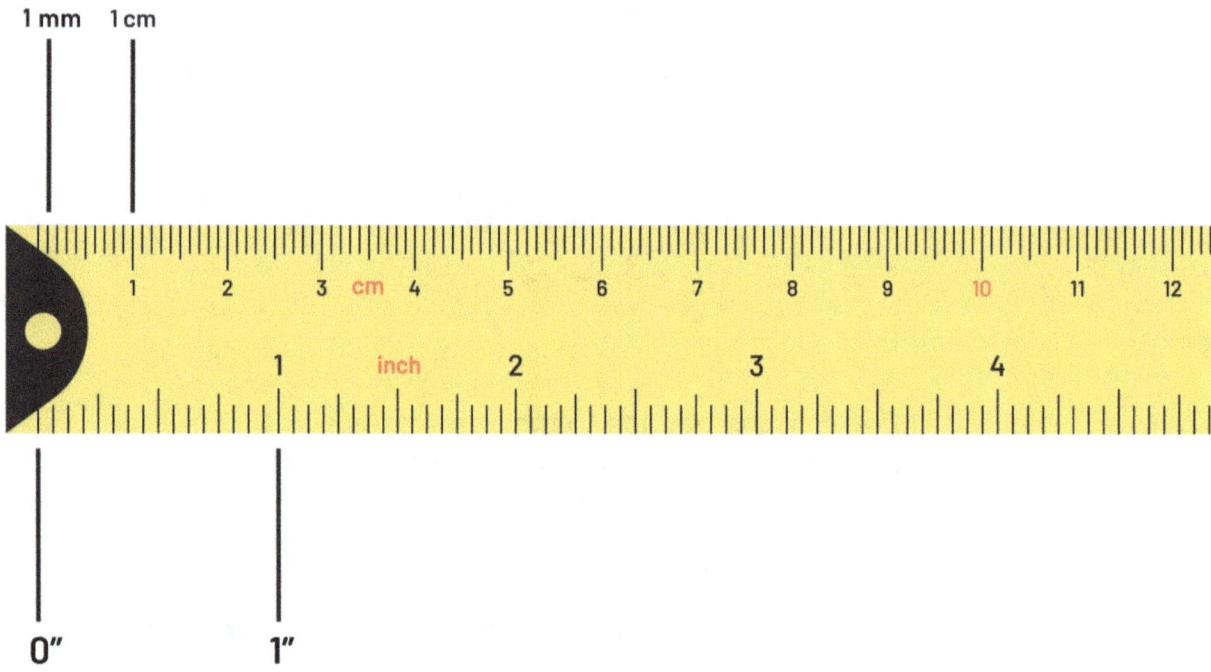

12 inches (in) = 1 foot

3 feet (ft) = 1 yard (yd)

5,280 feet = 1 mile

Teacher instructions: Show students a 12-inch ruler (which is 1 foot long) and a yard stick (which is 3 feet long).

One cubic foot is equal to how many cubic inches?
1,728 in³

Solution: 1 foot = 12 in.; $1^3 = 12^3$ (12 × 12 × 12 = **1,728 in³**)

One square yard is equal to how many square feet?
9 ft²

Solution: 1 yd. = 3 ft.; $1^2 = 3^2$ (3 × 3 = 9 ft²)

A piece of yarn x yards in length was cut into 4 congruent pieces. If the pieces were all 2.7 feet in length and no yarn was left over, what was the value of x?
(Hint: Don't forget that since 3 feet = 1 yard, you convert feet to yards by dividing by 3.)

x = 3.6 or $\frac{18}{5}$ yards

Solution: 4 × 2.7 = 10.8 ft.; 10.8 ÷ 3 = **3.6** or 18/5 yards
$\frac{36}{1} = \frac{36}{10} = \frac{18}{5}$

Note: To change a decimal answer such as 3.6 to a fraction, begin by recording 36 (the number without the decimal) as the numerator of the fraction and record the number 1 as the denominator (36/1). Since there is one number to the right of the decimal point in 3.6, add one zero after the number 1 to get 36/10. (If there were two numbers to the right of the decimal point, you would have recorded two zeros, and so forth.) Then reduce 36 over 10 to 18 over 5 by dividing both numbers by 2.

Trigonometry

Instruction: *Trigonometry* means triangle measurement; this branch of mathematics examines the ratios between the sides of right triangles. *Cos* stands for "cosine," *sin* stands for "sine," and *tan* stands for "tangent." You might need to find the cosine of angle A, the sine of angle A, or the tangent of angle A. For tan A below, you would record the length of the side opposite angle A as the numerator (6) and the length of the side adjacent to angle A as the denominator (8). Then you would reduce the ratio to lowest terms. (**Note:** A couple of ideas for learning the formulas are included on the following page.)

cos A = $\dfrac{\text{adjacent } \angle A}{\text{hypotenuse}}$	sin A = $\dfrac{\text{opposite } \angle A}{\text{hypotenuse}}$	tan A = $\dfrac{\text{opposite } \angle A}{\text{adjacent } \angle A}$

$$\tan A = \frac{6}{8} = \frac{3}{4} \text{ or } .75$$

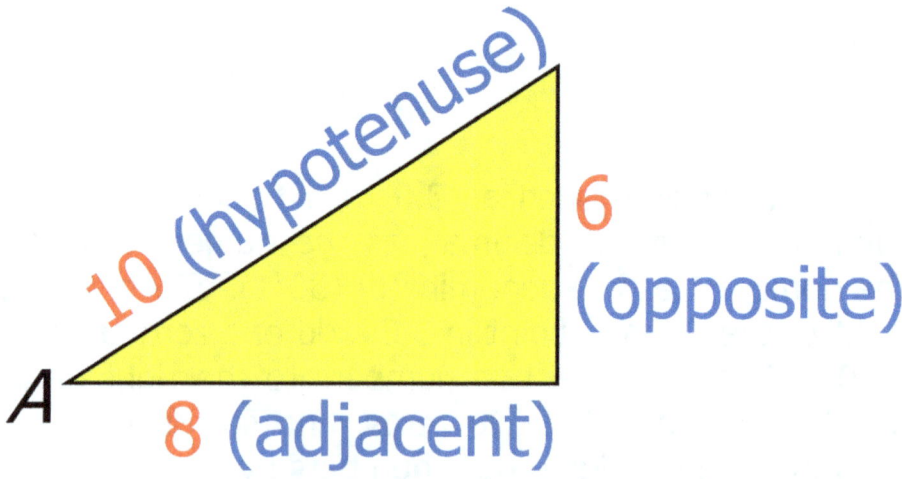

1st idea for learning the formulas on the previous page

Cosine (ah)   **Sin** (oh)   **Tangent** (oa)

1. **Cosine** (ah): Think of "ah" as something you would say if someone did something nice, such as if someone were to "cosign" for someone else, which is pronounced the same as "cosine." ("Ah" is the abbreviation for *adjacent over hypotenuse*.)

2. **Sin** (oh): The abbreviation of *sine* is *sin*, and "oh!" is something that might be said in response to sin. ("Oh" is the abbreviation for *opposite over hypotenuse*.)

3. **Tangent** (oa): Tangent is opposite over adjacent. To remember this, you might think of a toad, which starts with the same letter as *tangent*. The vowels in *toad* are *oa*, which can stand for *opposite over adjacent*.

2nd idea for learning the formulas on the previous page
   Sung to: "Froggie Went a Courtin'"

   Cosine equals adjacent over hypotenuse;
   sine equals opposite over hypotenuse;
   tangent equals op-po-site over ad-ja-cent;
   tangent equals opposite over adjacent.

Find cos A and sin A for ∠A. Then find tan C for ∠C. Reduce each ratio to lowest terms.

cos $A = \dfrac{8}{10} = \dfrac{4}{5}$ or .8

sin $A = \dfrac{6}{10} = \dfrac{3}{5}$ or .6

tan $C = \dfrac{8}{6} = \dfrac{4}{3}$ or 1.3

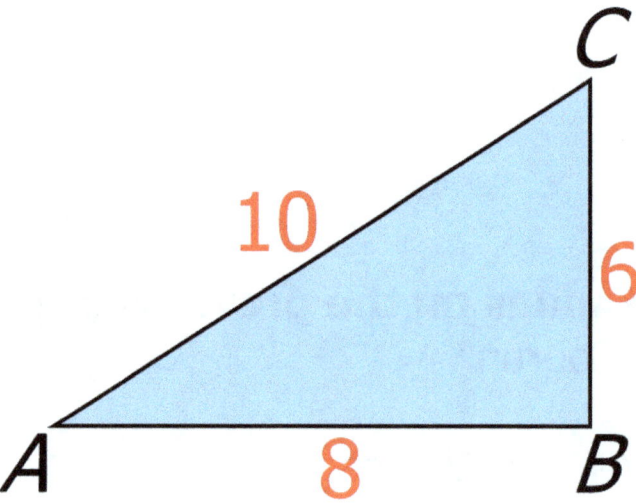

Find cosine B in the triangle below and use your calculator to reduce the ratio to lowest terms.

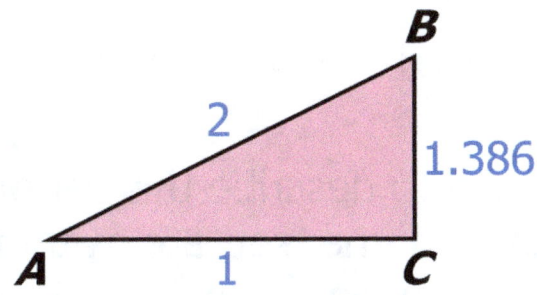

$\cos B = \dfrac{1.386}{2}$ or 0.693

Sets

list method
$V = \{$carrots, corn, peas$\}$

Instruction: The list method was used to describe this set of carrots, corn, and peas. The items in the set are within brackets, and a capital letter is used to name a set. The vegetables listed by name in this particular set are known as *elements* or *members* of the set.

set-builder notation
$V = \{x | x$ is a vegetable$\}$

Instruction: Set-builder notation was used to describe the set of carrots, corn, and peas. Set-builder notation starts out with $\{x|x$ is...$\}$ and then goes on to describe the set without listing the individual items in the set.

Set Notation

union

This is the symbol for *union*.

intersection

This is the symbol for *intersection*.

subset

This is the symbol for *subset*.

proper subset

This is the symbol for *proper subset*.

element of

Instruction: This is the symbol meaning "element of." If you put a slash through the symbol like ∉, it would mean "not an element of."

null set

Instruction: This is the symbol meaning "null set." The null set (or empty set) has zero elements. It is considered to be a subset of every set.

Instruction: Sets *A* and *D* below are **equal sets** because they have the exact same **elements**. Sets *A* and *B* are examples of **equivalent sets** because they have the same number of elements.

$A = \{\bullet, ⬠\}$ $C = \{⬠\}$

$B = \{\bullet, ■\}$ $D = \{⬠, \bullet\}$

$U = \{\bullet, ■, ⬠\}$

Instruction: This is the **universal set** in this example. Every shape from the four sets is listed once in the universal set.

$A \cup C$ $\bullet, ⬠$

Instruction: This example applies only to sets *A* and *C*. The **union** of *A* and *C* is the circle and pentagon. The shapes from both sets were listed, but the ⬠ was not listed twice.

$A \cap B$ $\bullet$

Instruction: The **intersection** of *a* and *b* is the circle; it is the shape that is in both sets.

C' $\bullet, ■$

Instruction: The **complement** of *C* is set *B* (or ●, ■) because it has all of the elements in the universal set that are not in set *C*.

Use sets A, B, C, D, and the universal set to answer the next two questions.

$$U = \{1, 2, 3, 4, 5, 6\}$$

$A = \{1, 5, 6\}$ $C = \{2, 3, 5\}$
$B = \{1, 2, 4, 6\}$ $D = \{1, 5\}$

D' $\{2, 3, 4, 6\}$

Instruction: This is everything from the universal set except for the numbers that are also in set D. To work this problem, list all the numbers from the universal set that are not in set D. Since there are several numbers in the universal set, it might be easier to just list the numbers from the universal set and then to cross out or erase the numbers that are in set D.

$(C \cup D)'$ $\{4, 6\}$

Instruction: This is every number from the universal set that is not in set C or in set D. You might just list the numbers from the universal set and then cross out or erase the numbers that are in set C and set D.

$A \cap B'$ $\{5\}$

Instruction: To work this problem, you would list any numbers from set A that are not in set B.

SUBSETS AND PROPER SUBSETS: Subsets and proper subsets are discussed on the next page. Think of a subset (or a proper subset) as being a part of that set. Is set A, C, or D above a subset or a proper subset of set B? The answer is no. However, if there was a set with just 2 and 4 in it $\{2, 4\}$, that set would be a subset and a proper subset of set B because set B has both 2 and 4 in it. On the other hand, if that set also had the number 8 in it $\{2, 4, 8\}$, it would not be a subset of B because 8 is not a part of B.

A = { 🔵, ⬟ }
B = { 🔵, 🟧 }
C = { ⬟ }
D = { ⬟, 🔵 }

$C \subseteq A$

$A \subseteq D$

Instruction: Set *C* is a **subset** of *A* because every element of set *C* is also an element of set *A*. (In the same way, set *C* is a subset of *D*.) A subset may also be equal, so *A* is a subset of set *D*. (Think of the line under the sideways *u* as being half of an equal sign.) Likewise, *A* is a subset of *A* ($A \subseteq A$), *B* is a subset of *B*, and so on. The **null set** (or empty set) is also a subset of *A* ($\emptyset \subseteq A$) because the null set is considered a subset of every set.

$C \subset A$

Instruction: Set *C* is a **proper subset** of *A* because every element of set *C* is also an element of set *A*. (Likewise, set *C* is a proper subset of *D*.) Note that proper subset does not allow for the sets to be equal; this is the difference between subset and proper subset. The symbol for proper subset does not have half an equal sign, so it cannot be equal. However, the **null set** (or empty set) is a subset of *A* ($\emptyset \subset A$).

Venn Diagram

Instruction: Notice the three sets listed below. If we wanted to make a Venn diagram to illustrate $R \cap B$ (set R intersects with set B), we would only be interested in the first two sets (sets R and B) and would ignore the last set (set Y). Notice that "1" is recorded in both the red and blue set, so we would record 1 on the Venn diagram where it is both red and blue (but no other colors); "2" and "4" are included in the red set but not the blue set, so we would record 2 and 4 on the Venn diagram where it is only red; "3" is recorded in the blue set but not the red set, so we would record 3 where it is only blue.

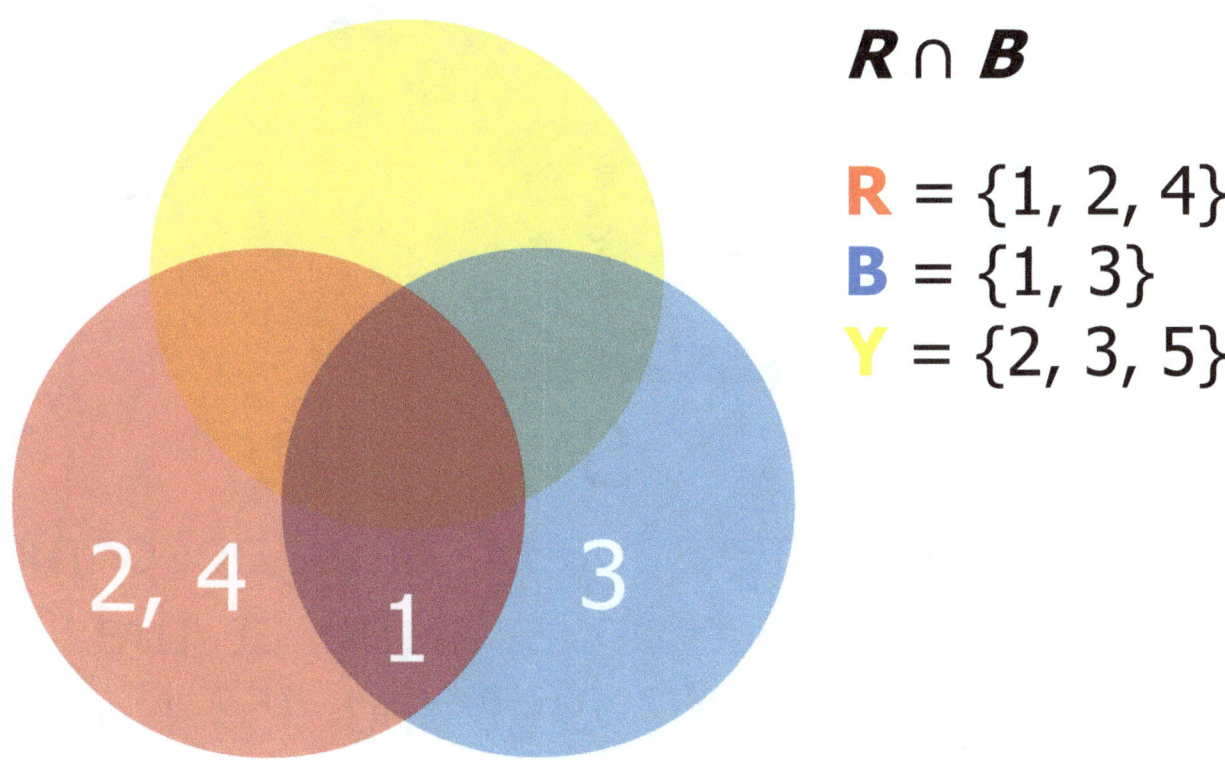

$R \cap B$

R = {1, 2, 4}
B = {1, 3}
Y = {2, 3, 5}

Practice: Use the Venn diagram to illustrate $B \cap Y$.

EQUATOR: Every globe is marked with two great circles. The equator is the circle on a globe that divides the earth into the Northern and Southern Hemispheres.

PRIME MERIDIAN: The prime meridian (and International Date Line) make up the great circle that divides the earth into the Eastern and Western Hemispheres.

Latitude: Latitude is the degree measured north or south of the equator; it is between 0° and 90°.

Longitude: Longitude is the degree measured east or west of the prime meridian; it is between 0° and 180°. The latitudinal and longitudinal lines that form a grid on a globe are used to pinpoint locations on the earth.

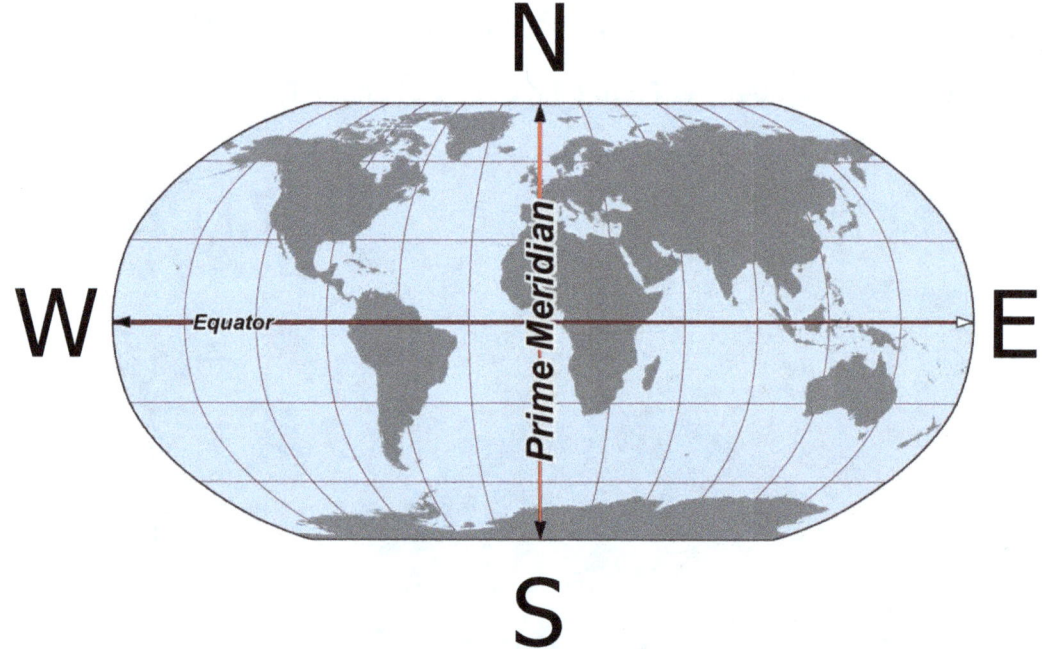

Begin where the equator and the prime meridian intersect on a globe at the 0° mark. Then go 90° East and 45° North. What continent is found at this location? Asia

Conditional $p \rightarrow q$	**Biconditional** $p \leftrightarrow q$
$T \rightarrow T = T$ $F \rightarrow F = T$ $F \rightarrow T = T$ $T \rightarrow F = F$	$T \leftrightarrow T = T$ $F \leftrightarrow F = T$ $F \leftrightarrow T = F$ $T \leftrightarrow F = F$
Instruction: A conditional statement takes the form "If p, then q." An example would be, "<u>If</u> Thanksgiving is in November, <u>then</u> Thanksgiving is in summer." Notice that a conditional statement is only false if p is true and q is false. In this example, p is "If Thanksgiving is in November" (which is true), and q is "then Thanksgiving is in summer" (which is false). Therefore, T + F = F.	**Instruction:** A biconditional statement takes the form "p if and only if q," such as, "Thanksgiving is in November <u>if and only if</u> Thanksgiving is in summer." Notice that for biconditional statements, p and q must both be true or both be false in order for the statement to be true. Thus, in this example, T + F = F.

converse

$$p \rightarrow q \quad q \rightarrow p$$

Instruction: The <u>converse</u> of $p \rightarrow q$ (if p, than q) is $\boldsymbol{q} \rightarrow \boldsymbol{p}$ (if q, then p); the q and p trade places.

inverse

$$p \rightarrow q \quad \sim p \rightarrow \sim q$$

Instruction: The <u>inverse</u> of $p \rightarrow q$ (if p, than q) is $\sim\boldsymbol{p} \rightarrow \sim\boldsymbol{q}$ (if not p, then not q); the symbol for "not" is placed before p and q. Note that the symbol $\neg$ is also used for "not."

contrapositive

$$p \rightarrow q \quad \sim q \rightarrow \sim p$$

Instruction: The <u>contrapositive</u> of $p \rightarrow q$ (if p, than q) is $\sim\boldsymbol{q} \rightarrow \sim\boldsymbol{p}$ (if not q, then not p), also written as $\neg\boldsymbol{q} \rightarrow \neg\boldsymbol{p}$. The q and p trade places, and the symbol for "not" is placed before q and p.

<u>If</u> it is snowing, <u>then</u> the temperature is below 33°.
   Converse: <u>If</u> the temperature is below 33°, <u>then</u> it is snowing.

   Inverse: <u>If</u> it is not snowing, <u>then</u> the temperature is not below 33°.

   Contrapositive: <u>If</u> the temperature is not below 33°, <u>then</u> it is not snowing.

Symbols

not (negation)

Instruction: Both symbols shown above can be used for "not" or "negation."

What is the negation (or opposite) of "Farmers grow corn"?
Farmers do <u>not</u> grow corn.

---

there exists	all

Instruction: This is the symbol for "there exists"; it is the *existential quantifier*. An example sentence is "<u>There exist</u> farmers who grow corn."

Instruction: This is the symbol for "all"; it is the *universal quantifier*. An example sentence is "<u>All</u> farmers grow corn" or "<u>Every</u> farmer grows corn."

Inductive Reasoning

Instruction: Inductive reasoning is used to argue that a statement is <u>most likely true</u>. See the example of this type of argument below.

❖ For the last 3 years, average test grades in Mrs. Clayton's class have ranged between 85% and 95%. Therefore, it is expected that the average test grades the following year will also range between 85% and 95%.

Although there is a good chance that the conclusion will happen, there are no guarantees. This type of argument is known as an ***appeal to tendency*** because it is relying upon what "tends" to happen or what has happened in the past to determine future results.

Deductive Reasoning

Instruction: Deductive reasoning is used to argue that a statement is <u>definitely true</u>. (See the example of this type of argument below.) A student once suggested that we think of "detective" when we hear the word *deductive*.

❖ Every quadrilateral has 4 sides, and every trapezoid is a quadrilateral. Therefore, every trapezoid has 4 sides.

In this argument, the **premises** are in red, and the **conclusion** is in blue.

Invalid and Valid Arguments

1. Every quadrilateral has 4 sides, and every trapezoid is a quadrilateral. Therefore, a right triangle has one right angle. (This sentence is **invalid** because the conclusion has nothing to do with the premises.)

2. Kailey has a Miniature Pinscher. A Miniature Pinscher took the frisbee. Therefore, Kailey's dog took the frisbee. (This sentence is **invalid** because "a Miniature Pinscher" isn't proof that it was Kailey's Miniature Pinscher.)

3. All children like spinach. Michael likes spinach. Therefore, Michael is a child.

    Instruction: The *major premise* says that all children like spinach, but it doesn't tell us that only children like spinach.

    Moreover, the *minor premise* tells us that Michael likes spinach, but it does not tell us that Michael is a child. Therefore, since the sentence concludes that he is a child just because he likes spinach, it is **invalid**.

    A **valid** argument might say: "All children like spinach. Michael is a child. Therefore, Michael likes spinach."
    or
    "Only children like spinach. Michael likes spinach. Therefore, Michael is a child."
    (This shows that a premise does not have to be true in order to be valid. Ridiculous sentences can be valid, but they cannot be *sound*.)

Sound and Unsound Arguments

A sound argument is a valid argument with true premises.

❖ Every quadrilateral has 4 sides, and every trapezoid is a quadrilateral. Therefore, every trapezoid has 4 sides.

Sound argument: This is a sound argument because it is valid, and the premises are true.

❖ A regular polygon is equilateral and equiangular. A scalene triangle is equilateral and equiangular. Therefore, a scalene triangle is a regular polygon.

Unsound argument: This argument is unsound. It is valid, but the second premise (*a scalene triangle is equilateral and equiangular*) is untrue.

Counterexample: If you can think of a counterexample, then you can prove that a conclusion is untrue. For example, if the conclusion of an argument is, "Therefore, every triangle has one right angle," the counterexamples *obtuse triangle* or *acute triangle* (neither of which have a right angle) would prove the conclusion false.

Note to the teacher: *This marks the end of Chapter 6. However, you are encouraged to review the next page with students. While it is taught in the elementary grades as well as in pre-algebra and algebra I in this program, students will also need to be familiar with the process of graphing ordered pairs in algebra II.*

Reviewing How to Plot Points on a Coordinate Plane

Instruction: When graphing an ordered pair, such as (2, −3), you start at the **origin**. (The black circle in the center of the **coordinate plane** below where the x-axis and y-axis intersect is the origin.) If the first number (called the x-coordinate) is positive 2, you would move to the right 2 units on the x-axis (if it were negative, you would move to the left). If the second number (called the y-coordinate) is negative 3, you would move down 3 units on the y-axis (if it were positive, you would move up). Thus, negative numbers are graphed to the left and down, while positive numbers are graphed to the right and up.

A (2, −3)

B (−4, 0)

C (0, 2)

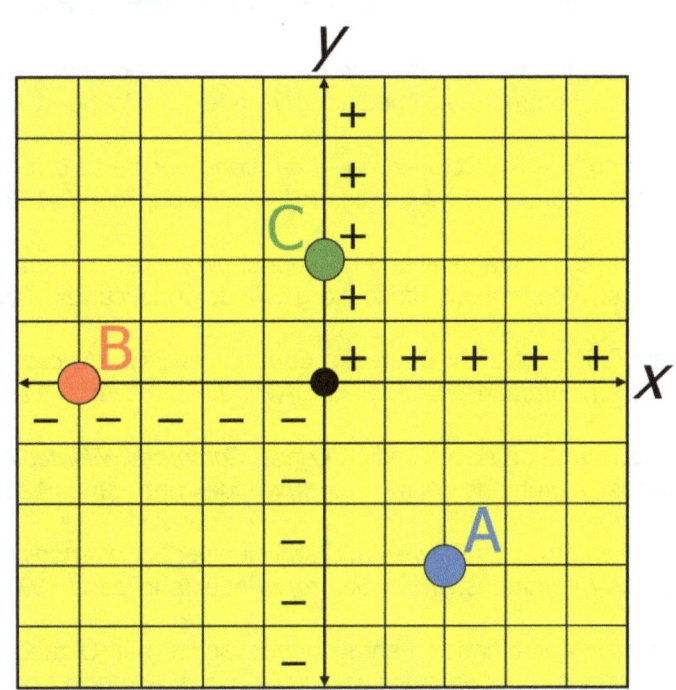

***A* (2, −3)**: Graph *A* by moving to the right 2 units and down 3 units.

***B* (−4, 0)**: Graph *B* by moving to the left 4 units. Do not move up or down since *y* is 0.

***C* (0, 2)**: To graph *C*, you do not move left or right because *x* is 0, but you do move up 2 units.

Additional practice: Record other ordered pairs for students to help graph on the classroom screen.

Amousey. Blank Venn Diagram. *Commons.wikimedia.org*. Wikimedia Commons. https://commons.wikimedia.org/w/index.php?curid=92368629.

Averater. Simple pyramid with height. *Commons.wikimedia.org*. Wikimedia Commons. https://commons.wikimedia.org/w/index.php?curid=22250216.

DaBler at cs.Wikipedia. Cuboid Simple. *Commons.wikimedia.org*. Wikimedia Commons. https://commons.wikimedia.org/w/index.php?curid=1945982.

DEMcAdams. Pentagonal Prism (Heptahedron). *Commons.wikimedia.org*. Wikimedia Commons. https://commons.wikimedia.org/w/index.php?curid=29499092.

DEMcAdams. Pentagonal Pyramid (Hexahedron). *Commons.wikimedia.org*. Wikimedia Commons. https://commons.wikimedia.org/w/index.php?curid=29499155.

Kmf164. Map showing the equator and Prime Meridian. *Commons.wikimedia.org*. Wikimedia Commons. https://commons.wikimedia.org/wiki/File:Primemeridian.jpg.

Murmann, Frank. Dodecahedron. *Commons.wikimedia.org*. Wikimedia Commons. https://commons.wikimedia.org/w/index.php?curid=15824469.

Murmann, Frank. Octahedron. *Commons.wikimedia.org*. Wikimedia Commons. https://commons.wikimedia.org/w/index.php?curid=15824531.

Roy, Mario. Pentagonal and hexagonal pyramids. *Commons.wikimedia.org*. Wikimedia Commons. https://commons.wikimedia.org/w/index.php?curid=10624986.

Scientif38. Protractor Degrees. *Commons.wikimedia.org*. Wikimedia Commons. https://commons.wikimedia.org/w/index.php?curid=12811448.

Sebbe.wigmo on sv.wikipedia. Cone. *Commons.wikimedia.org*. Wikimedia Commons. https://commons.wikimedia.org/w/index.php?curid=1266300.

Svdmolen at Dutch Wikipedia. Parallelepiped. *Commons.wikimedia.org*. Wikimedia Commons. https://commons.wikimedia.org/w/index.php?curid=1938463.

Tesserale_Kombination_Pentagondodekaeder_mit_Oktaeder_im_Gleichgewicht.png: FischX. Icosahedron. *Commons.wikimedia.org*. Wikimedia Commons. https://commons.wikimedia.org/w/index.php?curid=8343907.

Wersję rastrową wykonał użytkownik polskiego projektu wikipedii: Siałababamak, Zwektoryzował: Krzysztof Zajączkowski. Different cones-diagrams. *Commons.wikimedia.org*. Wikimedia Commons. https://commons.wikimedia.org/w/index.php?curid=11105455.

Ævar Arnfjörð Bjarmason. Cylinder (geometry). *Commons.wikimedia.org*. Wikimedia Commons. https://commons.wikimedia.org/w/index.php?curid=776995.

A2569875. Cuboid. *Commons.wikimedia.org*. Wikimedia Commons. https://commons.wikimedia.org/w/index.php?curid=82634886.

www.ingramcontent.com/pod-product-compliance
Lightning Source LLC
Chambersburg PA
CBHW080547230426
43663CB00015B/2740